my revision notes

T0187394

Edexcel International GCSE (9–1)

BIOLOGY

Nick Dixon

Boost

HODDER
EDUCATION
AN HACHETTE UK COMPANY

The publisher would like to thank the following for permission to reproduce copyright material:

Photo credits: p. 1 Kletr - Fotolia; p.2 Ramona Heim – Fotolia; p.25 © egal – istock via Thinkstock; p.47 royaltystockphoto/123RF.com; p.83 © Leonid Andronoc – Fotolia; p.91 © Dr Jeremy Burgess/Science Photo Library; p.105 © Nigel Cattlin/Alamy; p.115t defun – Fotolia; p.115b © eugenesergeev – Fotolia.com

Although every effort has been made to ensure that website addresses are correct at time of going to press, Hodder Education cannot be held responsible for the content of any website mentioned. It is sometimes possible to find a relocated web page by typing in the address of the home page for a website in the URL window of your browser.

Hachette UK's policy is to use papers that are natural, renewable and recyclable products and made from wood grown in well-managed forests and other controlled sources. The logging and manufacturing processes are expected to conform to the environmental regulations of the country of origin.

Orders: please contact Hachette UK Distribution, Hely Hutchinson Centre, Milton Road, Didcot, Oxfordshire, OX11 7HH. Telephone: +44 (0)1235 827827. Email education@hachette.co.uk. Lines are open from 9 a.m. to 5 p.m., Monday to Friday. You can also order through our website: www.hoddereducation.co.uk

© Nick Dixon 2018

First published in 2018 by
Hodder Education
An Hachette UK Company,
Carmelite House, 50 Victoria Embankment
London EC4Y 0DZ

Impression number   5   4
Year   2024

Cover photo © fivespots – stock.adobe.com

Illustrations by Aptara, Inc

Typeset in India by Aptara, Inc

Printed in CPI Group (UK) Ltd, Croydon, CR0 4YY

ISBN 9781510446731

Get the most from this book

Everyone has to decide his or her own revision strategy, but it is essential to review your work, learn it and test your understanding. These Revision Notes will help you to do that in a planned way, topic by topic. Use this book as the cornerstone of your revision and don't hesitate to write in it — personalise your notes and check your progress by ticking off each section as you revise.

Tick to track your progress

Use the revision planner on pages iv and v to plan your revision, topic by topic. Tick each box when you have:

- revised and understood a topic
- tested yourself
- practised the exam practice questions and gone online to check your answers and complete the quick quizzes.

You can also keep track of your revision by ticking off each topic heading in the book. You may find it helpful to add your own notes as you work through each topic.

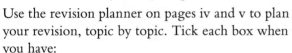

Features to help you succeed

Exam tips

Expert tips are given throughout the book to help you polish your exam technique in order to maximise your chances in the exam.

Typical mistakes

The author identifies the typical mistakes candidates make and explains how you can avoid them.

Now test yourself

These short, knowledge-based questions provide the first step in testing your learning. Answers are at the back of the book.

Definitions and key words

Clear, concise definitions of essential key terms are provided where they first appear.

Key words from the specification are highlighted in bold throughout the book.

Revision activities

These activities will help you to understand each topic in an interactive way.

Exam practice

Practice exam questions are provided for each topic. Use them to consolidate your revision and practise your exam skills.

Summaries

The summaries provide a quick-check bullet list for each topic.

Online

Go online to check your answers to the exam practice questions and try out the extra quick quizzes at **www.hoddereducation.co.uk/ myrevisionnotesdownloads**

My revision planner

REVISED   TESTED   EXAM READY

Now test yourself answers

Exam practice answers and quick quizzes at
www.hoddereducation.co.uk/myrevisionnotesdownloads

Countdown to my exams

6–8 weeks to go

- Start by looking at the specification — make sure you know exactly what material you need to revise and the style of the examination. Use the revision planner on pages iv and v to familiarise yourself with the topics.
- Organise your notes, making sure you have covered everything on the specification. The revision planner will help you to group your notes into topics.
- Work out a realistic revision plan that will allow you time for relaxation. Set aside days and times for all the subjects that you need to study, and stick to your timetable.
- Set yourself sensible targets. Break your revision down into focused sessions of around 40 minutes, divided by breaks. These Revision Notes organise the basic facts into short, memorable sections to make revising easier.

REVISED ☐

2–6 weeks to go

- Read through the relevant sections of this book and refer to the exam tips, exam summaries, typical mistakes and key terms. Tick off the topics as you feel confident about them. Highlight those topics you find difficult and look at them again in detail.
- Test your understanding of each topic by working through the 'Now test yourself' questions in the book. Look up the answers at the back of the book.
- Make a note of any problem areas as you revise, and ask your teacher to go over these in class.
- Look at past papers. They are one of the best ways to revise and practise your exam skills. Write or prepare planned answers to the exam practice questions provided in this book. Check your answers online and try out the extra quick quizzes at **www.hoddereducation.co.uk/myrevisionnotesdownloads**
- Use the revision activities to try out different revision methods. For example, you can make notes using mind maps, spider diagrams or flash cards.
- Track your progress using the revision planner and give yourself a reward when you have achieved your target.

REVISED ☐

One week to go

- Try to fit in at least one more timed practice of an entire past paper and seek feedback from your teacher, comparing your work closely with the mark scheme.
- Check the revision planner to make sure you haven't missed out any topics. Brush up on any areas of difficulty by talking them over with a friend or getting help from your teacher.
- Attend any revision classes put on by your teacher. Remember, he or she is an expert at preparing people for examinations.

REVISED ☐

The day before the examination

- Flick through these Revision Notes for useful reminders, for example the exam tips, exam summaries, typical mistakes and key terms.
- Check the time and place of your examination.
- Make sure you have everything you need — extra pens and pencils, tissues, a watch, bottled water, sweets.
- Allow some time to relax and have an early night to ensure you are fresh and alert for the examinations.

REVISED ☐

My exams

Edexcel International GCSE (9–1) Biology paper 1B

Date:..

Time: ....

Location: ....

Edexcel International GCSE (9–1) Biology paper 2B

Date:..

Time: ....

Location: ....

1 Living organisms: variety and common features

Characteristics of living organisms

All living organisms share the following characteristics:
- They require nutrition to provide energy.
- They respire to release energy.
- They excrete their waste.
- They respond to their surroundings.
- They move.
- They control their internal conditions (often to respond to changes in their environment).
- They reproduce.
- They grow and develop.

Anything that completes all of these life processes is alive. Animals, plants, fungi and bacteria complete these processes and so are considered living. Viruses do not complete all of these processes (for example, they do not require nutrition or respire), so they are non-living.

> ### Now test yourself
> TESTED
>
> 1 What does nutrition provide?
> 2 What process releases energy?
> 3 Why might living organisms control their internal conditions?
> 4 Why are viruses not considered to be alive?
>
> Answers on p. 124

Variety of living organisms

Eukaryotic organisms

REVISED

Eukaryotic organisms, or eukaryotes, have cells with a nucleus. Animal, plant, fungal and protoctist cells are eukaryotic. Humans are eukaryotes, so almost all our cells have a nucleus containing our DNA.

> DNA (deoxyribonucleic acid): The genetic information found in all living organisms.
>
> Multicellular: Made from more than one cell.

Plants

Plants are multicellular organisms. Their cells contain chloroplasts. Chloroplasts are small, green cellular components that contain chlorophyll. Photosynthesis occurs in this green chloroplast pigment. This process stores energy from light as glucose, which then allows the plant to grow. Plants also store excess glucose as starch or sucrose. Unlike the cells of animals, plant cells are surrounded by a cellulose cell wall, which provides additional protection and support. Examples of plants include flowering plants such as cereals like maize, and herbaceous legumes such as peas or beans.

Animals

Animals are also multicellular organisms. They do not possess chloroplasts and so are unable to photosynthesise. This means they must

Figure 1.1 **A mosquito insect sucking blood from a human. Both mosquitos and humans are classed as animals.**

consume plants or other animals for food. Their cells are only surrounded by membranes (and not walls). Animals usually have a nervous system for coordination, and move from one place to another to find mates and food, or to avoid being hunted. They often store excess glucose as glycogen. Examples of animals include mammals (like humans) and insects like mosquitos and houseflies.

Fungi

Fungi can be either single-celled, like yeast, or multicellular, like mushrooms. Their cell walls are made from **chitin**. Fungi cannot photosynthesise. They feed by secreting digestive **enzymes** outside of their body (extracellular secretion) onto their food (which is often dead and decaying matter). They then absorb the digested organic products of their broken-down food. This is called **saprotrophic** nutrition. They may store any excess glucose as glycogen.

Another example of a fungus is *Mucor*. *Mucor* has a typical fungal structure that many others also possess. Its body is organised into a **mycelium** made from thread-like structures called **hyphae**, which often contain many nuclei.

Protoctists

Protoctists are microscopic, single-celled organisms. Some protoctists, like *Amoeba*, have characteristics of animals. *Amoebae* live in pond water. They are able to make bulges in their cytoplasm called pseudopods. This ability allows them to change shape, move and take in food. Other protoctists, like *Chlorella*, are more similar to plants than animals. They have chloroplasts containing chlorophyll and so can photosynthesise.

Some protoctists cause disease – they are **pathogenic**. **Malaria** is passed from person to person by mosquitos, but it is actually the protoctist *Plasmodium* that causes the disease. This protoctist is passed from person to person by the mosquito in the blood it drinks. Malaria causes recurring fevers, which can be fatal if not treated. The spread of malaria can be reduced by preventing mosquitos from breeding, and also by using mosquito nets and sprays to avoid being bitten.

Prokaryotic organisms

REVISED

Prokaryotic organisms, or **prokaryotes**, have cells without a nucleus. They do have a cell wall, cell membrane and cytoplasm. Bacterial cells do not have a nucleus and so bacteria are prokaryotes. All bacteria are single-celled and are microscopic – they are usually smaller than eukaryotic cells. Bacterial cells have a circle of chromosomal DNA within their cytoplasm. The cytoplasm also contains small rings of DNA called plasmids. Some bacteria can photosynthesise, but most feed from other organisms.

Examples of bacteria include *Lactobacillus bulgaricus*, which is a rod-shaped bacterium used to convert milk into yoghurt, and *Pneumococcus*, which is a pathogenic, spherical bacterium that causes **pneumonia**.

Pathogenic bacteria may produce poisons (toxins) that damage tissues and make us ill. However, not all bacteria cause disease. Many, including those in your digestive system, are useful to us. Bacteria live alongside

Figure 1.2 The *Mucor* fungus with its characteristic hyphae

Chitin: A polymer made from sugars that forms the cell walls of fungi and the exoskeletons of insects.

Enzyme: A biological molecule that speeds up a chemical reaction.

Saprotrophic: A type of consumer that obtains its nutrition from dead or decaying organisms by extracellular excretion of enzymes.

Mycelium: The vegetative part of a fungus, made of many branching, thread-like hyphae.

Hyphae: The branching, thread-like filaments which make up many fungi.

Pathogenic: Something that causes disease.

Malaria: A communicable disease, caused by a protoctist and transmitted by mosquitos, which attacks red blood cells.

Prokaryote: A microscopic, single-celled organism without a nucleus. Bacteria are examples of prokaryotes.

Pneumonia: A lung inflammation in which a bacterial (or viral) infection causes the build-up of pus in the lungs.

other living organisms and are often found in places like our mouths, noses and throats. They do not live inside individual cells like viruses.

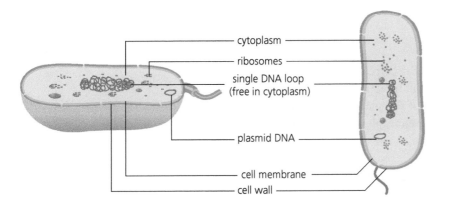

Figure 1.3 **A bacterial cell shown in three- and two-dimensional diagrams.**

The components of prokaryotic cells and their functions are shown in Table 1.1.

Table 1.1 **Components and functions of prokaryotic cells**

Component	Structure and function
Cytoplasm	This fluid is part of the cell inside the cell membrane. It is mainly water and it holds other components such as ribosomes. This is where most of the chemical reactions in the cell happen (such as the making of proteins in ribosomes).
Cell wall	Like plant and fungi cells, bacterial cells have a cell wall to provide support. However, unlike plant cell walls they are not made of cellulose. A cell membrane is found on the inside surface of the cell wall.
Single DNA loop (no chromosomes)	DNA in prokaryotes is not arranged in chromosomes, as it is in eukaryotic cells. Instead it is found in a single loop within the cytoplasm. Prokaryotes do not have a nucleus.
Plasmids	These are small, circular sections of DNA found in the cytoplasm of bacteria. They provide **genetic variation** for bacteria.
Cell membrane	This controls what substances go into and out of a cell. It also has internal extensions that have enzymes attached to them. This is where respiration occurs.
Ribosomes	Ribosomes are found in the cytoplasm. They make proteins.

Revision activity

Make a list of all the characteristics of eukaryotic and prokaryotic cells on separate flash cards. Then arrange them into two lists, one for each type of organism, to check you understand the differences between them.

Revision activity

Draw out Table 1.1 with only the components and headings listed. Try to fill the rest of the table in from memory to help you revise.

Typical mistake

Don't make the mistake of thinking that bacterial DNA is arranged in chromosomes. It is actually arranged as a single loop in the cytoplasm.

Genetic variation: Differences within a species of organism that can be inherited.

Viruses

REVISED

Viruses are not alive because they do not complete all of the characteristics of living things (see page 1). For example, they replicate instead of reproducing, and they do not respire. Because they are not alive, viruses are not classed as living species, but instead are **strains**. Viruses are small particles, even smaller than bacteria. They exist in a variety of shapes and sizes, and are made from a short length of nucleic acid (DNA or RNA) surrounded by a protein coat.

Strain: A type of virus.

Viruses are parasites and can only reproduce inside another cell. When they reproduce, they cause the host cell to burst, which releases new viruses to infect surrounding cells. Viruses can infect every type of living organism.

The tobacco mosaic virus

An example of a virus is the tobacco mosaic virus. This virus causes discolouring of tobacco plants because the infection stops the formation of chloroplasts. This reduces the ability of the plant to photosynthesise and therefore limits its growth. Another example of a common virus is the influenza virus, which causes the 'flu'.

HIV (human immunodeficiency virus)

HIV is a virus spread by exchange of bodily fluids. This can occur through sexual contact or when blood is swapped in shared needles used by drug users. Initially an infected person will feel flu-like symptoms. Unless an infected person is given antiviral drugs, the virus will attack their body's immune system. HIV turns into AIDS (acquired immune deficiency syndrome) when the person's immune system can no longer defend them.

Pathogens

REVISED

Pathogens are micro-organisms that pass disease from one organism to another. Diseases that can be spread by pathogens are called communicable (or infectious) diseases. There are four types of pathogen:

- viruses such as HIV
- bacteria such as salmonella
- fungi such as rose black spot in plants
- protoctists such as malaria.

The lifecycle of all pathogens is similar:
1 They infect a host.
2 They reproduce (or replicate if a virus).
3 They spread from their host.
4 They infect another host and the lifecycle repeats.

Spread of pathogens

REVISED

Pathogens spread disease in different ways, as shown in Table 1.2.

Vector: An animal that transmits a communicable disease.

Table 1.2 **Methods by which different pathogens spread disease**

Method of distribution	Example
Airborne	The virus that causes the common cold can be spread through the air in tiny droplets when people sneeze.
Direct contact (sexual or non-sexual)	Chlamydia is a sexually-transmitted disease (STD) that passes from one person to another during sex.
Dirty water	Cholera is caused by a bacterium that is spread in dirty water.
Contaminated food	Food poisoning is often caused by the *Escherichia coli* bacterium in undercooked or reheated food.
Vectors	Some farmers believe that the tuberculosis bacterium is passed from badgers to their cows. Animals that spread pathogens like this are called **vectors**.

The spread of disease can be reduced or prevented by:
- high levels of personal hygiene (for example correct hand washing)
- covering your mouth and nose when you cough or sneeze
- cleaning and disinfecting surfaces and objects with antiseptics
- being vaccinated and taking medicines as prescribed
- avoiding close contact with people who are sick.

Levels of organisation

REVISED

Some fungi, all bacteria and most protoctists are single-celled, whilst all animals and plants are multicellular. Larger multicellular organisms are more complex than single-celled organisms and have often evolved several levels of organisation within their bodies. These levels of organisation are listed below from smallest to largest:
- Organelles are the structures that make up cells. Examples include mitochondria and ribosomes.
- Cells are the basic building blocks of all life.
- Tissues are groups of cells with similar structures and functions.
- Organs are groups of tissues that perform specific functions.
- Organ systems are groups of organs with similar functions.
- Organisms are made from various organ systems.

Table 1.3 **Examples of levels of organisation**

Organisation level	Examples
Organelle	Mitochondrion, ribosome
Cell	Nerve cell, muscle cell
Tissue	Nervous tissue, tendons
Organ	Brain, heart
Organ system	Nervous system, digestive system
Organism	Human, frog

Revision activity

Draw out Table 1.3 with only the headings. Try to fill the rest of the table in from memory to ensure that you know the content.

Now test yourself

TESTED

5 Explain the difference between prokaryotes and eukaryotes.
6 Name a substance that animals use to store energy.
7 Name a major group of eukaryotes that can be either single-celled or multicellular.
8 What is saprophytic nutrition?
9 What is the polymer that forms the cell walls of fungi?
10 In what way is the protoctist *Chlorella* like a plant?
11 What are pathogens?
12 What are plasmids?
13 Define the term organelle.
14 Put the following in order of size starting with the smallest: organ systems, organisms, cells, organs, tissues, organelles.

Answers on p. 124

Cells and their organisation

Generalised cells

Generalised animal and plant cells have components in common with bacterial cells, as described on page 2. They have cytoplasm, in which most chemical reactions occur, as well as a cell membrane that controls what enters and exits the cell. A generalised animal cell is seen in Figure 1.4 and a generalised plant cell in Figure 1.5.

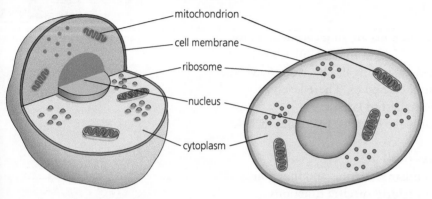

Figure 1.4 **A generalised animal cell as three- and two-dimensional diagrams**

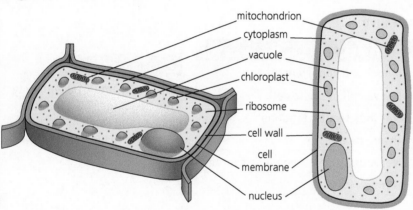

Figure 1.5 **A generalised plant cell as three- and two-dimensional diagrams**

Generalised plant cells have all the same components as animal cells: a nucleus, cytoplasm, cell membrane, mitochondria and ribosomes. Many plant cells also have a vacuole, which is a store of sugary cell sap, and chloroplasts, where photosynthesis occurs. They also have a cell wall to provide structure.

The functions of the additional components of animal and plant cells are shown in Table 1.4.

Table 1.4 The components present in generalised animal and plant cells and their functions

Component	Present in:		Function
	Animal cells	Plant cells	
Nucleus	Yes	Yes	Contains the DNA (genetic information) of an organism arranged into chromosomes.
Cytoplasm	Yes	Yes	Fluid part of the cell inside the cell membrane. It is mainly water and holds other components such as ribosomes. This is where most of the chemical reactions in the cell happen (such as the making of proteins in ribosomes).
Cell membrane	Yes	Yes	Controls which substances go in and out of a cell.
Mitochondria	Yes	Yes	Organelles found in the cytoplasm that release energy from glucose by respiration.
Ribosomes	Yes	Yes	Organelles found in the cytoplasm where proteins are made.
Vacuole	No	Yes	Plant cell structure that contains cell sap. This is where dissolved sugars and mineral ions are stored.
Chloroplasts	No	Yes	Contain chlorophyll, which is where photosynthesis occurs.
Cell wall	No	Yes	Used to provide support in plant cells, as in bacteria and fungi. Plant cell walls are made from cellulose.

Exam tip

It can be easy to confuse the components of generalised plant, animal and bacterial cells but it is important to remember which components are in each one. You must also be able to link the structures of each to their functions.

Mineral ions: Substances that are essential for healthy plant growth, such as nitrates and magnesium.

Typical mistake

Don't confuse the nucleus of a cell with the nucleus of an atom (in chemistry). It is important to remember that the nucleus of a cell is much larger than the nucleus of an atom. The nucleus of a cell is made of millions of atoms, each with their own atomic nucleus.

Revision activity

You could make flash cards of generalised plant, animal and bacterial cells. On one side draw a labelled diagram and on the other write the components and their functions.

Cell differentiation

REVISED

Complex multicellular organisms like humans, other animals and plants are not made of just one type of cell. There are around 200 different types of cell in each human being. Each cell type has become specialised for a specific function. This process is called differentiation. Cells can develop specific components during this process.

Most animal cells become specialised, or differentiated, before birth. Unlike animal cells, many plant cells retain their ability to differentiate throughout their entire life. For example, if you take a plant cutting of a small section of stem and place it in soil, new differentiated root cells are still able to grow.

Differentiation: The biological process of developing into something different or specialised.

Typical mistake

Students often just state **how** a specialised cell is adapted, rather than explaining **why** it is adapted. It is important to check what the question is asking and answer it correctly.

Examples of differentiated cells

REVISED

Some examples of important differentiated cells are described below.

Sperm cells

Sperm cells have a tail to propel them towards the **ovum** (egg). For their very small size, they have many **mitochondria** to enable them to release energy during respiration. Their nucleus contains DNA from the father, which will make up half of the DNA of the new organism.

Nerve cells

Nerve cells (neurones) pass electrical signals around your body to control and coordinate your actions. They have a long **axon**, along which the electrical signals quickly pass. This is insulated by the **myelin sheath**. They have branching nerve endings that can communicate with hundreds of surrounding cells.

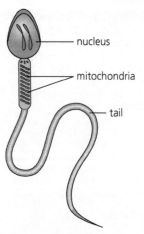

Figure 1.6 **The parts of a sperm cell**

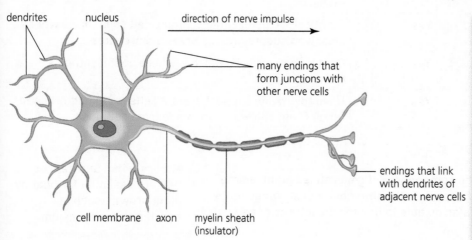

Figure 1.7 **The parts of a nerve cell**

Muscle cells

Muscle cells contract and relax to move parts of our bodies. This can be involuntary (automatic) like your heartbeat, or voluntary like moving your fingers to type an email. They have large numbers of mitochondria to release the energy from glucose in respiration.

Ova (plural), ovum (singular): Egg(s)

Mitochondrion (plural: mitochondria): A cell organelle where respiration occurs; found in the cytoplasm of eukaryotic cells.

Axon: The extension of a nerve cell along which electrical impulses travel.

Myelin sheath: The insulating cover along an axon which speeds up the electrical signal.

Revision activity

You could make flash cards of specialised cells. On one side you could draw a labelled diagram and on the other side you could write the functions.

Exam tip

For the exam, you should be able to explain how the structure of specialised cells relates to their function. It is not enough to just say what their structure or function is – you need to know how they both link.

Stem cells

REVISED

Stem cells can develop into other types of cell. Without stem cells, no multicellular organisms could exist.

Stem cells in mammals

There are two types of stem cell found in humans and other mammals. During fertilisation, your mother's ovum and father's sperm fused to

Stem cell: An undifferentiated cell that can develop into one or more types of specialised cell.

form one embryonic stem cell. This cell then divided (by mitosis) until you were fully formed. For the first nine weeks of your life in the womb, your cells were not specialised and had not differentiated. They remained embryonic stem cells. These cells could develop into any of the 200 different cell types you possess. Cells with this ability are called totipotent cells.

The second type of stem cell found in humans and other mammals is the adult stem cell. Despite the name, adult stem cells start to develop much earlier than when you become an adult. Adult stem cells are found in specific locations in your body like bone marrow and your nose. They can only develop into one or two cell types. For example, adult stem cells in your bone marrow can develop into blood cells, whilst those in your nose can develop into nerve cells. These are called multipotent cells.

Some animals have stem cells that can differentiate into many types even as adults. Lizards can shed their tails, if caught by a predator, and then regrow them. The ability to regrow the tail is due to their stem cells. Another example is a starfish – if one of its limbs is severed, the severed limb will grow four new ones, and the original starfish will grow one new limb, giving two fully-formed individual starfish.

Stem cells and differentiation in plants

Plant stem cells are also found in specific locations called meristems. These regions are in the tips of shoots and roots. Much of a plant's growth occurs in these regions. Unlike your adult stem cells, plant stem cells retain the ability to differentiate throughout their life. This means we can take a cutting of a small plant stem and place it in soil to form a clone, because its stem cells will start to develop into roots.

Stem cell research

Stem cell research uses stem cells to develop medical treatments that, in the future, could treat patients with damaged cells. The idea is that by transplanting new cells into injured or non-working organs, the damage might be repaired. It is hoped that this could be used to rebuild a damaged spinal cord, or treat the brains of people with Parkinson's disease or the pancreases of diabetes patients. Totipotent embryonic stem cells are more useful in research than adult stem cells because they can develop into all types of cell.

Using a person's own stem cells in medical treatments means their bodies are far less likely to reject the new cells than if they were transplanted from another person. The process of making an embryo with the same genes as the parent is called therapeutic cloning. However, there is a small possibility of transferring viral infections in this process.

Stem cell research involves ethical issues. This means some people disagree with it for religious or moral reasons. Some people donate unused fertilised ova from in vitro fertilisation for stem cell research. The use of these cells is controversial because there are questions over whether they are a life. Because of these issues, tight regulations surround all scientific studies involving stem cells.

Embryonic: Something that has an embryo state, as in the early stages of development. It usually refers to an unborn human child.

Totipotent: Describes a stem cell that can develop into any type of specialised cell.

Multipotent: Describes a stem cell that can only develop into a few particular types of specialised cell.

Meristem: An area of a plant (normally the tip of a root or shoot) in which rapid cell division occurs.

Parkinson's disease: A medical condition that reduces brain function, usually seen in patients of advanced years.

Diabetes: A non-communicable disease that reduces a person's control of their blood glucose concentrations. This can lead to a coma if not carefully monitored and treated.

Ethical issue: An idea some people disagree with for religious or moral reasons.

In vitro fertilisation (IVF): A medical procedure in which an ovum is fertilised outside of a woman, then placed into her uterus to develop into a baby.

Revision activity

Stem cell research is an ethical issue. Research a list of arguments for and against this process.

Now test yourself

TESTED

15 State all of the cell components that are present in plant cells but not in animal cells.
16 What process happens in mitochondria?
17 What is the name given to the part of a neurone along which electrical impulses travel?
18 What is the difference between adult and embryonic stem cells?
19 Define the term totipotent.
20 What name is given to an area of a plant in which rapid cell division occurs?
21 In what regions of a plant are its meristems?
22 What is the difference between the stem cells found in plants and those found in animals?
23 What is an ethical issue?
24 Define the term *in vitro* fertilisation.

Answers on p. 124

> **Exam tip**
>
> Students often forget that stem cells are present in plants as well as animals. It is important that you can describe the function of stem cells in embryos, adult animals and plant meristems.

Biological molecules

Carbohydrates, proteins and lipids

REVISED

Carbohydrates

Carbohydrates are biological molecules that contain carbon, hydrogen and oxygen. They are made from simple sugars. The smallest carbohydrates are **monosaccharide** sugars such as glucose and **fructose**. **Disaccharides** are larger sugars and include **sucrose** and **lactose**.

Starch is an even larger carbohydrate. It is made from large numbers of glucose molecules joined together. Starch is the carbohydrate that most plants use to store excess energy. It is very common in our diet and is found in foods like potatoes, wheat, corn and rice.

Glycogen is a carbohydrate that is similar in structure to starch. It is used as a store of energy in animals, fungi and bacteria.

Cellulose is another larger carbohydrate made from hundreds or thousands of glucose molecules. Cellulose forms the cell walls of plants.

> **Monosaccharide:** A simple sugar that cannot be broken down into smaller ones.
>
> **Fructose:** A simple sugar found in fruit and honey.
>
> **Disaccharide:** A larger sugar made from two monosaccharide sugars.
>
> **Sucrose:** A simple sugar that we use in cooking.
>
> **Lactose:** A simple sugar found in milk.
>
> **Starch:** A large carbohydrate made from glucose that is used as a store of energy in plants.
>
> **Glycogen:** A large carbohydrate made from glucose that is used as a store of energy in animals, fungi and bacteria.
>
> **Cellulose:** A large carbohydrate made from glucose that is used to make cell walls in plants.

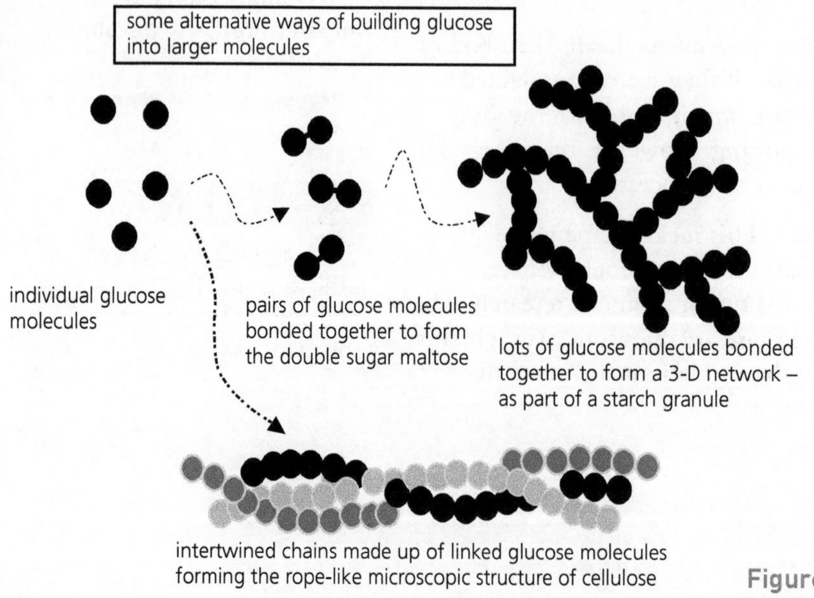

some alternative ways of building glucose into larger molecules

individual glucose molecules

pairs of glucose molecules bonded together to form the double sugar maltose

lots of glucose molecules bonded together to form a 3-D network – as part of a starch granule

intertwined chains made up of linked glucose molecules forming the rope-like microscopic structure of cellulose

Figure 1.8 **The structure of a carbohydrate**

Proteins

Proteins are large biological molecules made from long chains of **amino acids**. They are formed in a process called protein synthesis, which occurs in **ribosomes**. Examples of proteins include hormones and enzymes. Proteins have many functions within living organisms and they are vital to many of the metabolic processes that keep us alive.

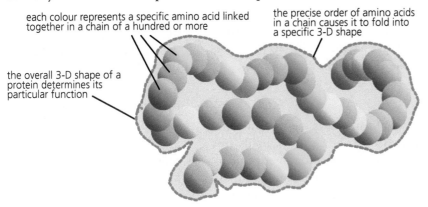

each colour represents a specific amino acid linked together in a chain of a hundred or more

the precise order of amino acids in a chain causes it to fold into a specific 3-D shape

the overall 3-D shape of a protein determines its particular function

Figure 1.9 **The structure of a protein**

Lipids

Lipids are fats and oils. Fats are solid at room temperature, whilst oils are liquid. Lipids are mainly used as a chemical store of energy. Lipids are made from three fatty acid chains and one glycerol molecule.

> **Amino acids:** The components of proteins, which are assembled into the correct sequence and shape during protein synthesis (see page 82).
>
> **Ribosomes:** Cellular components found within the cytoplasm in which proteins are made.
>
> **Lipids:** Fats or oils, which are insoluble in water.

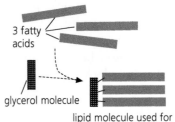

3 fatty acids

glycerol molecule

lipid molecule used for storing energy inside cells

Figure 1.10 **The structure of a lipid**

Required practical 1

Investigate food samples for the presence of starch, glucose, protein and fat

Equipment:
- Spotting tile
- Iodine solution
- Boiling tubes
- Distilled water
- Benedict's solution
- Water bath
- Biuret solution
- Bung

Method and results of starch test:
- A small amount of food was placed onto a spotting tile.
- Two drops of iodine solution were added.
- If the food turned blue or black, starch was present.
- If it remained brown (the colour of iodine solution) then no starch was present.

Method and results of glucose test:
- A small amount of food was placed in a boiling tube.
- 10 cm³ of distilled water was added.
- Ten drops of Benedict's solution were added to the boiling tube.
- The boiling tube was heated in a water bath at 90 °C for ten minutes.
- If the solution turned orange or green, glucose was present.
- If it remained blue (the colour of Benedict's solution) then no glucose was present.

Method and results of protein test:
- A small amount of food was placed in a boiling tube.
- 10 cm³ of distilled water was added.
- Ten drops of Biuret solution were added to the boiling tube.
- If the solution turned a light lilac colour, protein was present.
- If it remained blue (the colour of Biuret solution) then no protein was present.

Method and results of oils test:
- A small amount of food was placed in a boiling tube.
- 10 cm³ of distilled water was added.
- A bung was placed in the boiling tube and it was shaken vigorously.
- If oil was present, an emulsion formed and the water turned cloudy.

Enzymes

Enzymes are biological **catalysts** that speed up reactions but are not used up in them. To better understand enzymes and their role, we can look at the example of enzymes in our digestive systems. These enzymes break down large molecules of food into smaller ones and so are called breakdown enzymes.

Some enzymes do the opposite and join smaller molecules together to make larger ones. These are called synthesis enzymes. The enzymes involved in protein synthesis are examples of synthesis enzymes.

The lock and key theory

Enzymes are specific to their **substrates**, just like keys are specific to their locks. For example, protease enzymes won't break down lipids, just as the key to your house won't open your car. In order for an enzyme to break down a substrate, the substrate must fit into the enzyme, just like a key fits into a lock. The shapes of the enzyme and substrate match just like keys and locks, which is why we call it the **lock and key theory**.

> **Enzyme:** A biological molecule that speeds up a chemical reaction.
>
> **Catalyst:** A substance that increases the rate of a reaction without being used up itself.
>
> **Substrate:** The molecule or molecules that an enzyme acts upon.
>
> **Lock and key theory:** A model that explains the specific action of enzymes.

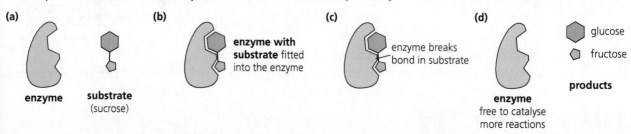

Figure 1.11 How a digestive enzyme breaks down a substrate. Here the substrate is sucrose and the products are glucose and fructose.

At optimum pH and temperature, the shape of the enzyme and substrate fit perfectly. As the conditions move away from the optimum pH or temperature, the shape of the **active site** changes. This change makes it harder for the enzyme and substrate to fit together and so slows the rate at which the enzyme works, which slows the reaction. If extremes of pH or temperature are reached, the shape of the active site is permanently changed. The enzyme will become **denatured** and will no longer function.

> **Active site:** The region of an enzyme that binds to its substrate.
>
> **Denatured:** Describes an enzyme that has been permanently changed by extremes of pH or temperature, which stop it from working.
>
> **Kinetic energy:** A store of movement energy.

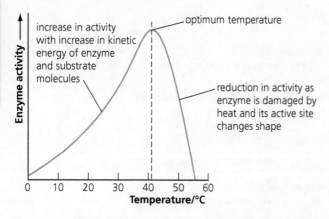

Figure 1.12 Graph showing the effect of temperature on the activity of an enzyme

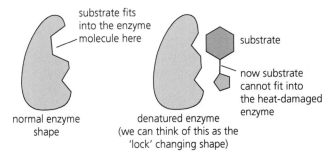

normal enzyme shape

denatured enzyme (we can think of this as the 'lock' changing shape)

substrate fits into the enzyme molecule here

substrate

now substrate cannot fit into the heat-damaged enzyme

Figure 1.13 **Extremes of temperature and pH denature enzymes by altering the shape of their active site, so the substrate no longer fits.**

Exam tip

Students often struggle to relate the activity of enzymes to their temperature. Make sure you can describe why the rate of reaction slows at low temperatures and stops at high ones. Enzyme activity at different temperatures can be explained by particle theory. At higher temperatures, molecules have more **kinetic energy** so move faster. This means they are more likely to collide with substrates. At very high temperatures, however, the enzyme will be denatured.

Required practical 2

Investigate how enzyme activity can be affected by changes in pH

Equipment:
- Starch solution
- Boiling tubes
- Iodine solution
- Amylase solution
- pH buffer solutions
- Water bath
- Pipettes
- Spotting tile

Method:
- $10\,cm^3$ of starch solution was placed in a boiling tube.
- Ten drops of iodine solution were added.
- $2\,cm^3$ of amylase solution was added to another boiling tube.
- $5\,cm^3$ of pH buffer solution was added to the second boiling tube.
- Both boiling tubes were placed into a water bath at 37 °C for two minutes.
- One drop of iodine was placed into each well of a spotting tile.
- After two minutes, the contents of both boiling tubes were mixed together.
- Every 30 seconds, a pipette was used to put one drop of solution into a new well of the spotting tile. This was repeated until the starch test no longer worked; that is, the solution remained brown and didn't turn blue or black.
- This was repeated with different pH buffer solutions.

Results:
- The longer the iodine test gave positive results by turning blue or black, the more suitable the pH was for the enzyme.
- The optimum pH was that which gave the most positive results.

Required practical 3

Investigate how enzyme function can be affected by changes in temperature

Equipment:
- Starch solution
- Boiling tubes
- Iodine solution
- Amylase solution
- pH buffer solutions
- Water bath
- Pipettes
- Spotting tile

Method:
- $10\,cm^3$ of starch solution was placed in a boiling tube.
- Ten drops of iodine solution were added.
- $2\,cm^3$ of amylase solution was added to another boiling tube.
- Both boiling tubes were placed into a water bath at 37 °C for two minutes.

→

- One drop of iodine was placed into each well of a spotting tile.
- After two minutes, the contents of both boiling tubes were mixed together.
- Every 30 seconds, a pipette was used to put one drop of solution into a new well of the spotting tile.
- This was repeated until the starch test no longer worked; that is, the solution remained brown and didn't turn blue or black.
- This was repeated at different temperatures.

Results:
- The longer the iodine test gave positive results by turning blue or black, the more suitable the temperature was for the enzyme.
- The optimum temperature was that which gave the most positive results. This was 37 °C.
- At 100 °C the enzyme denatured. No reaction occurred and so the solution remained brown.
- At 20 °C similar results were seen to the optimum temperature, but they were slower.
- At 0 °C, no reaction occurred and so the solution remained brown.

Now test yourself

TESTED ☐

25 In which substances is fructose sugar found?
26 What are starch and cellulose made from?
27 Hormones and enzymes are types of which biological molecule?
28 What are proteins made from?
29 Where in a cell are proteins made?
30 What is the difference between fats and oils?
31 Describe the food test for starch.
32 What fits into the active site of an enzyme?
33 Describe what occurs when an enzyme is denatured.
34 Which theory explains how enzymes are specific to their substrates?

Answers on p. 124

Exam tip

When asked about the activity of enzymes, you should remember how enzyme activity relates to changes in pH and include it in your answer.

Movement of substances into and out of cells

Diffusion

REVISED ☐

Diffusion is the spreading out of particles resulting in the net (overall) movement of particles from an area of higher to lower concentration. This happens naturally and does not require energy, so we call it a **passive process**. Because diffusion always happens from higher to lower concentration, we say it occurs **down** a **concentration gradient**.

Exam tip

We use the term 'net' to mean overall movement of particles by diffusion (and osmosis) because a smaller number of particles may move backwards (from low to high concentration). This is why it is important to use the correct terminology in the exam.

Particles of gases and liquids can diffuse. Those of solids have fixed positions and so cannot.

Diffusion: The net movement of particles from an area of higher concentration to an area of lower concentration.

Passive process: A process that occurs naturally without the need for energy.

Concentration gradient: A measurement of how the concentration of a substance changes from one place to another.

Examples of diffusion

Some common examples of diffusion within organisms are outlined below.

Oxygen in your lungs

Diffusion occurs in many places in your body. One example is in your lungs, where you breathe oxygen into the **alveoli** found there. In this example oxygen diffuses from a higher concentration within your alveoli to a lower concentration within your red blood cells. When these blood cells absorb oxygen, their oxygen concentration increases from low to higher. When the blood cells move through your blood vessels, oxygen moves from the high concentration in your blood cells to the lower concentration in your body cells. The body cells have a low concentration of oxygen because they have been respiring and using it up.

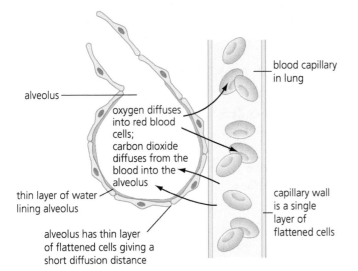

Figure 1.15 **Diffusion of gases between an alveolus and a blood capillary in the lung**

This whole process is then repeated in reverse with carbon dioxide, which is produced during respiration by your body's cells. This carbon dioxide dissolves straight into your **blood plasma**, rather than being absorbed by the red blood cells as oxygen is.

Glucose in your digestive system

Another example of diffusion occurs in your digestive system. Sugars like glucose are produced here when carbohydrates are broken down, so glucose is at a high concentration in your small intestine. The glucose will therefore move by diffusion from your small intestine to the lower concentration in your blood through villi (see page 27). When your blood absorbs the glucose, the blood gains a higher concentration. As the blood travels, this glucose then moves to your body's cells, which have a lower concentration because they have been respiring and using glucose up.

Urea in your cells

A final example of diffusion involves waste products. Some of your cells make **urea** as a waste product. This diffuses from a high concentration in the cells to a lower concentration in the blood. The urea is transported to the kidney where it is **excreted**.

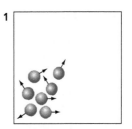

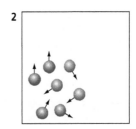

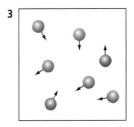

Figure 1.14 **Molecules in a gas spread out by diffusion.**

Alveoli (singular: alveolus): Tiny air sacs found in the lungs through which gases exchange between blood and air.

Blood plasma: The straw-coloured liquid that carries our blood cells and dissolved molecules.

Urea: A waste product of protein metabolism in mammals, which is excreted in urine.

Excretion: The removal of substances from cells or organisms.

Diffusion in other organisms

Your lungs are extremely effective at absorbing oxygen, but not all animals have complex exchange surfaces like these. Insects simply have small tubes that run into their bodies, into which gases diffuse. This system is much less sophisticated than your lungs, and so insects cannot complete as much gas exchange as humans. This limits their overall size. Their maximum size is in part determined by the distance that oxygen can diffuse easily into their cells.

Factors that affect diffusion

The factors that affect diffusion are shown in Table 1.5.

Table 1.5 **The factors that affect diffusion**

Factor	How it affects diffusion
Difference in concentration	If the two concentrations are similar, the rate of diffusion will be slow. The more different the two concentrations are, the quicker the rate of diffusion.
Temperature	At higher temperatures, all particles have more kinetic energy. The higher the temperature, the quicker the rate of diffusion.
Surface area	More diffusion can occur over a larger surface area. The larger the surface area, the quicker the rate of diffusion.

Smaller organisms like insects or single-celled organisms can survive without specialised exchange surfaces because they have a large surface area to volume ratio compared to larger organisms.

Typical mistake

Students often mix up how the three factors in Table 1.5 affect diffusion. Make sure you can correctly remember and explain how an increase and decrease in each example will affect diffusion.

Exam tip

Exams will often ask about the differences between exchange surfaces and transport systems, as well as why and when they might be needed. It's important to revise this topic carefully.

Revision activity

Draw and label a set of diagrams to explain how increases and decreases in the three factors shown in Table 1.5 affect diffusion.

Required practical 4

Investigate diffusion using non-living systems

Equipment:
- Agar jelly
- Phenolphthalein
- Dilute sodium hydroxide
- Dilute hydrochloric acid
- Beaker
- Knife

Method:
- Agar jelly was made with phenolphthalein and dilute sodium hydroxide mixed into it.
- The phenolphthalein stain coloured the jelly pink.
- The agar was cut into different sized cubes: 1 cm by 1 cm, 2 cm by 2 cm and 3 cm by 3 cm.
- The cubes were added to the beaker of hydrochloric acid.
- The time taken for the hydrochloric acid to diffuse into the jelly cubes was recorded. This turned the cubes from pink to colourless.
- The time taken for each cube size was compared.
- The surface area to volume ratio for each cube size was calculated.

Results:
- The rate of diffusion was calculated by dividing the time taken for the entire cube to turn colourless by the distance from the outside to the centre of the cube.
- Diffusion took longer in the larger cubes.

Osmosis

REVISED

Osmosis is the spreading out of water particles, resulting in the net movement of water from an area of higher to lower water concentration across a **partially permeable** membrane. So, osmosis is the diffusion of water across a membrane.

This happens naturally and does not require energy, so we call it a passive process. Because osmosis always happens from a higher to a lower concentration, we say it occurs **down** a concentration gradient.

Examples of osmosis

When it rains, the soil becomes wet and so it has a high concentration of water particles. This high water concentration in the soil is often higher than that in a plant's root hair cells, so water moves by osmosis into the plant from a higher to a lower concentration across the membrane of the root hair cells.

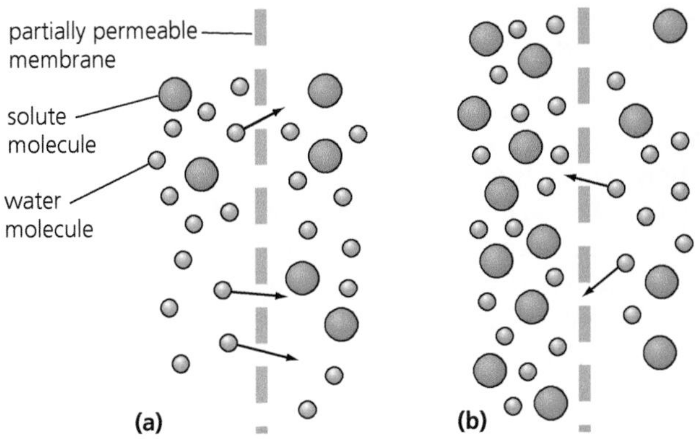

partially permeable membrane

solute molecule

water molecule

(a)   (b)

Figure 1.16 Two partially permeable membranes. Both membranes have the same concentration of solution on the right-hand side. However, water moves in opposite directions through the membranes.

We can see osmosis when we put cells into different solutions (Figure 1.17). In Figure 1.17a, the red blood cells and the plasma they are surrounded by have the same concentration. So, though osmosis does occur, with water moving from the cells into the plasma and from the plasma into the cells, no **net** movement occurs. Solutions with the same overall concentration are called **isotonic**.

If a red blood cell is put into a solution of salty water (as in Figure 1.17b), it will shrivel and shrink. This is because there is a higher concentration of water in the cell than that in the surrounding salty solution. The water will therefore move by osmosis into the solution. If one solution has a higher concentration than another, we call the one with the higher concentration **hypertonic**.

If a red blood cell is put into a solution of distilled water (as in Figure 1.17c), it will swell and may burst. This is because there is a lower concentration of water in the cell than in the surrounding solution. Water will move by osmosis from the solution into the cell. If a solution has a lower concentration than another one, we call the one with the lower concentration **hypotonic**.

(a)   (b)   (c)

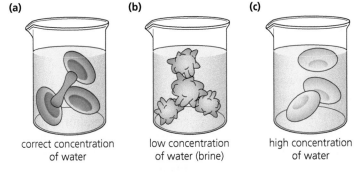

correct concentration of water

low concentration of water (brine)

high concentration of water

Figure 1.17 This is what happens to red blood cells in solutions with different concentrations of water. (Not to scale)

Required practical 5

Investigate osmosis using a living system

Equipment:
- Boiling tubes
- Five salt or sugar solutions
- Distilled water
- Potato disks

Method:
- Six small disks of potato were cut and the mass of each was recorded.
- Each disk was placed in a boiling tube containing a different salt or sugar solution.
- One disk was placed in a boiling tube of distilled water to act as a control.
- After 30 minutes, each disk was removed and dried on a paper towel before its mass was recorded.
- The percentage change in mass for each disk was calculated.

Results:
- A graph of concentration of salt or sugar solution and change in mass was plotted.
- If the solution contained more sugar or salt than the potato disk, then the potato disk lost mass. This was because water moved from a higher concentration in the disk to a lower concentration in the solution.
- If the solution contained less sugar or salt than the potato disk, then the potato disk gained mass. This was because water moved from a higher concentration in the solution to a lower concentration in the disk.
- If the mass of the disk stayed the same, then we can say that the concentration in the potato disk is the same as that of the solution and that no net movement occurred.

Active transport

REVISED

Active transport is the net movement of particles from an area of lower concentration to an area of higher concentration. So, active transport reverses the effects of diffusion.

This process does not happen naturally and so it requires the use of energy. It is therefore not passive like diffusion or osmosis, but an active process. Because active transport always happens from a lower to a higher concentration, we say it occurs **up** a concentration gradient.

> **Active transport:** The net movement of particles from an area of lower concentration to an area of higher concentration using energy.

Examples of active transport

Plant minerals

Water moves from a higher concentration in the soil to a lower concentration in a plant by osmosis. This does not require energy as it is a passive process. However, plants also need to absorb mineral ions from the soil. These minerals are present at higher concentrations in the plant than the soil, so they cannot move into the plant by diffusion. Instead plants need to use active transport, which uses energy.

Sugar in the digestive system

Active transport also occurs in your digestive system. When you have just digested a meal (Figure 1.18a), glucose is found at a high concentration inside your small intestine. This glucose moves naturally by diffusion into your blood and so this does not require energy. However, once most of this glucose has been absorbed, the glucose is now at a higher concentration in your blood and so cannot be absorbed any further by diffusion (Figure 1.18b). This means your body must absorb the last of the glucose into your blood by active transport, which uses energy.

(a)

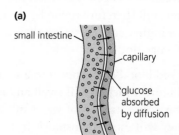

(b)

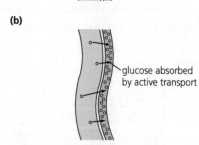

Figure 1.18 **Look carefully at the nutrient concentrations in the intestine and the blood: (a) diffusion and (b) active transport**

Now test yourself

TESTED ☐

35 Define the term diffusion.
36 Name two places where oxygen diffuses in the human body.
37 Define the term osmosis.
38 What term do we give to a solution with a lower concentration of solutes than a rival solution?
39 By what process does water move into plant root hair cells?
40 Define the term active transport.
41 What does active transport require that diffusion does not?
42 What two processes help move glucose from the small intestine into the capillaries?
43 Why do plants absorb mineral ions from the soil by active transport?

Answers on p. 124

Summary

- All living organisms share the following characteristics: they require nutrition to provide energy, they respire to release energy, they excrete their waste, they respond to the surroundings, they move, they control their internal conditions (often to respond to changes in their environment), they reproduce, and they grow and develop.
- Animal, plant, fungal and protoctist cells are eukaryotic, so they have a nucleus. Bacterial cells are prokaryotic and so do not.
- Fungi can be single-celled or multicellular. They feed by excreting enzymes outside of their bodies and then ingesting the broken-down food. They have a chitin cell wall.
- Most protoctists are single-celled microscopic organisms. Some, like *Chlorella*, can photosynthesise and so are more like plants. Others, like *Amoeba*, have characteristics of animals.
- Bacterial cells are smaller than plant or animal cells. They have cytoplasm, a cell membrane and a cell wall. Their DNA is found in a single loop. They may also have DNA plasmids.

- Viruses are not alive because they do not complete the eight characteristics of life. They are very small. They are made from a short length of DNA or RNA surrounded by a protein coat.
- Complex multicellular organisms are arranged into organ systems, then organs, then tissues, then cells and finally organelles.
- Cell specialisation allows cells to complete a specific function. Examples in animals include sperm, nerve and muscle cells.
- As organisms develop, their cells differentiate to become specialised. The cells develop different components to fulfil their functions. Differentiation happens early in the life of most animals. Many plant cells can differentiate throughout their life.
- Stem cells are undifferentiated. In humans, embryonic stem cells can develop into many different cell types, whilst adult stem cells can only develop into a few.
- Plant stem cells are found in regions called meristems and can develop into any cell type. They can be used to produce clones of rare plants or important crops.

→

- Carbohydrates are made from simple sugars. Proteins are made from amino acids. Lipids are made from fatty acids and glycerol.
- Enzymes are biological catalysts that speed up reactions. An enzyme's substrate fits into its active site. Enzymes are specific for their substrate. The lock and key theory models this. Enzymes have an optimum pH and temperature where their activity is greatest. At extremes of pH or temperature, the shape of the active site changes and the enzyme becomes denatured. The substrate will no longer fit.
- Diffusion is the net movement of particles from an area of higher to lower concentration.
- Osmosis is the net movement of water from a higher to a lower water concentration across a partially permeable membrane.
- Active transport is the net movement of particles from an area of lower to higher concentration and requires energy.

Exam practice

1 Cells are the basic units of life. All living organisms on Earth are made from one or more cells.
  (a) Which of the following is present in animal and plant cells but not bacterial ones? [1]
      A Cytoplasm
      B Cell membrane
      C Nucleus
      D Cellulose cell wall
  (b) Use the three terms listed below to copy and complete the table so that they have the correct definitions. [3]

Prokaryote   Mitochondrion   Pathogen

	A cell organelle in which respiration occurs.
	An organism that does not possess nuclei in its cell or cells (bacteria)
	A disease-causing micro-organism

  (c) The diagram on the right shows a generalised plant cell. Name the cell components A to C. [3]
  (d) Explain **two** differences between human embryonic and adult stem cells. [2]
  (e) Describe **four** similarities between plant cells and animal cells, and **two** differences. [6]

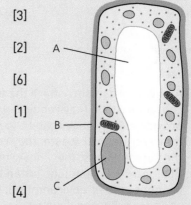

2 Food tests allow us to see which food groups are in the things we eat.
  (a) Which of the following reagents is used to test for protein? [1]
      A Benedict's
      B Iodine
      C Universal indicator
      D Biuret
  (b) Describe the procedures used to test for glucose and starch. [4]

3 Substances like digested food and oxygen move between cells. Without the processes that result in this movement, our cells would not receive all they need to survive.
  (a) Compare and contrast the processes of diffusion and active transport. [6]
  (b) Another transport process used by organisms is osmosis. Describe how you would investigate osmosis in the living system of a potato. [6]

4 Enzymes speed up reactions within cells. The graph shows how an enzyme was affected by different temperatures.
  (a) Describe how the enzyme's effectiveness changed in relation to the temperatures shown. [4]
  (b) Describe how you would investigate the effect of pH on the enzyme activity of carbohydrase enzymes. [6]

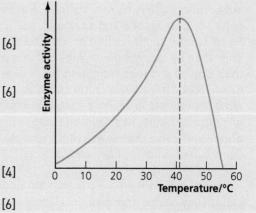

Answers and quick quizzes online

ONLINE

2 Nutrition and respiration

Nutrition in flowering plants

Photosynthesis

Plants take in water through their roots and carbon dioxide through their leaves. These reactants (water and carbon dioxide) are then converted into glucose during **photosynthesis**. Oxygen is created as a by-product of this reaction. Energy from light is stored in glucose as chemical energy during photosynthesis. The word equation for this reaction is:

$$\text{carbon dioxide + water} \xrightarrow{\text{light in}} \text{glucose + oxygen}$$

The balanced chemical symbol equation for photosynthesis is:

$$6CO_2 + 6H_2O \xrightarrow{\text{light in}} C_6H_{12}O_6 + 6O_2$$

Photosynthesis only occurs when light is present. This is because it is an **endothermic reaction** that requires energy, which is provided by light. Photosynthesis occurs in a green compound called **chlorophyll**, which is present in sub-cellular structures called **chloroplasts**. Chloroplasts are found in high numbers in **palisade mesophyll cells**, which are located in the top layers of leaves.

Plants need mineral ions to keep them healthy and allow them to grow, just like humans need vitamins and minerals. Plants absorb these minerals by active transport from the soil through their root hair cells (see page 18). Examples of the minerals they need are:
- tiny amounts of magnesium ions to make chlorophyll for photosynthesis
- nitrate ions to make amino acids for protein synthesis. These proteins are required for growth and repair.

> **Typical mistake**
>
> A typical mistake is to say that plants photosynthesise to make oxygen for humans and other animals. Oxygen is just a by-product of photosynthesis. Plants only photosynthesise in order to make glucose for themselves.

Photosynthetic algae

As well as plants, algae can also photosynthesise. Some algae are single-celled whilst others, like seaweed, are larger. Some algae have green chlorophyll like plants, but others use different coloured photosynthetic pigments. These are seen in brown and red seaweeds, for example. More than two-thirds of the oxygen made every day is by photosynthetic algae in our oceans rather than plants on land.

> **Photosynthesis**: A chemical reaction that occurs in the chloroplasts of plants and algae, which allows energy to be stored in glucose.
>
> **Endothermic reaction**: A reaction in which energy is absorbed from the surroundings.
>
> **Chlorophyll**: A green pigment found in chloroplasts that carries out photosynthesis.
>
> **Chloroplast**: Small, green structures found in green plant cells. They contain chlorophyll for photosynthesis.
>
> **Palisade mesophyll cells**: Cells found at the top of plant leaves, which contain high numbers of chloroplasts to maximise photosynthesis.

> **Exam tip**
>
> In both the word and symbol equations, you can write the two reactants and the two products either way around. However, you must not mix up the reactants and products as this would be incorrect.

> **Typical mistake**
>
> Don't forget that the cells of all living things (including plants) respire at all times. A typical mistake that students make is to say plants just photosynthesise and animals respire.

Structure of the leaf

REVISED

Leaves are plant organs, and different plants have evolved a large number of different shapes and sizes of leaf. Many have a similar structure and similar characteristics. For example, all leaves have a flat surface, called the blade, and a midrib, which runs down the middle from the stem to the tip of the leaf. The leaf's veins branch out from this midrib. Almost all leaves are adapted to absorb sunlight for photosynthesis, and they often have a large surface area to maximise this absorption. Many leaves are also thin, so that there is a shorter distance for gases to diffuse in and out of their cells.

Epidermal tissue

The outside layer of a plant is called its **epidermis**. This layer of cells has many functions including protection against water loss, exchange of gases between the leaves and the air, and also uptake of water in the roots. The epidermis is transparent to allow light to pass through it for photosynthesis.

Palisade mesophyll

Below the epidermis in leaves is the palisade mesophyll layer. This layer contains cells with very high numbers of chloroplasts to maximise the amount of glucose produced in photosynthesis.

Spongy mesophyll

Below the palisade mesophyll layer of leaves is the **spongy mesophyll** layer. The cells in this layer have fewer chloroplasts because they are further from the sunlight, so less photosynthesis happens here. The cells of this layer are less regularly shaped than in the palisade mesophyll layer and so they have gaps between them. These gaps allow gases to diffuse from within the leaf into the air and vice versa.

Stomata

Leaves have small pores called **stomata**, which are mainly found on the lower surface. They open and close to allow gases, including water vapour, to diffuse in and out. Stomata are surrounded by guard cells, which swell to open the stoma and shrink to close it.

> **Epidermis:** The outermost layer of cells in an organism.
>
> **Spongy mesophyll:** Cells found towards the bottom of leaves with spaces between them to allow gases to diffuse.
>
> **Stomata (singular stoma):** The tiny holes in leaves that are bordered by guard cells. These stomata allow gases to diffuse in and out of the leaf as the stomata open.

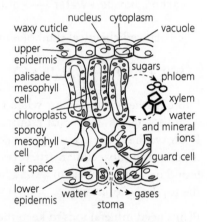

Figure 2.1 A cross-section of a leaf

Limiting factors

REVISED

The rate of a reaction is how quickly the reaction occurs. The rate at which plants and algae photosynthesise decreases when:
- temperatures fall (as the particles involved have less kinetic energy)
- carbon dioxide levels drop
- light intensities reduce (see below)
- plants do not have enough chlorophyll.

If one or more of these conditions occurs, the rate of photosynthesis becomes limited. These problematic conditions are then called '**limiting factors**'. Farmers can reduce the effects of limiting factors by keeping their crops in greenhouses or polytunnels. Within these they can:
- raise the temperature
- add carbon dioxide (commonly by adding a burner)
- provide artificial light.

Photosynthesis provides plants and algae with chemical energy in the form of glucose. They then use this glucose in the six key ways shown in Figure 2.2.

> **Limiting factor:** Anything that reduces the rate of a reaction or stops the reaction.

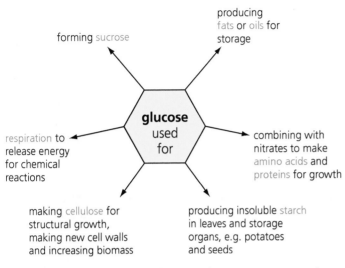

Figure 2.2 How plants make use of the glucose produced in photosynthesis

forming sucrose

producing fats or oils for storage

glucose used for

respiration to release energy for chemical reactions

combining with nitrates to make amino acids and proteins for growth

making cellulose for structural growth, making new cell walls and increasing biomass

producing insoluble starch in leaves and storage organs, e.g. potatoes and seeds

Plants or algae performing photosynthesis are present at the start of almost every food chain on our planet. This means that the process of photosynthesis provides the energy that supports almost all life on Earth, including your own. Without photosynthesis, there would be very little life on our planet.

> **Exam tip**
>
> Students often confuse the effects of temperature, light intensity, carbon dioxide concentration and amount of chlorophyll on the rate of photosynthesis. It is important that you can describe what happens when each of these factors increases and decreases.

Required practical 1

Investigate photosynthesis, showing the evolution of oxygen from a water plant

Equipment:
- Lamp
- Metre ruler
- Boiling tube
- Pondweed
- Water
- Splint

Method:
- The equipment was set up as shown in Figure 2.3.
- The boiling tube with pondweed was placed 10 cm from the lamp.
- The light was switched on and left for two minutes so the plant could acclimatise.
- The number of bubbles of oxygen given off in one minute was then recorded.
- The boiling tube was then moved further away in increments of 10 cm from the lamp and the experiment was repeated up to a maximum distance of 50 cm.

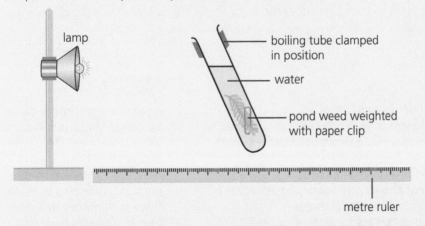

lamp

boiling tube clamped in position

water

pond weed weighted with paper clip

metre ruler

Figure 2.3 The equipment used to investigate the effect of light intensity on the rate of photosynthesis

Results:
- The results of the experiment produced the data in Figure 2.4.
- A glowing splint was added to the tube to see if oxygen was being produced. The splint relit, which proved the presence of oxygen.

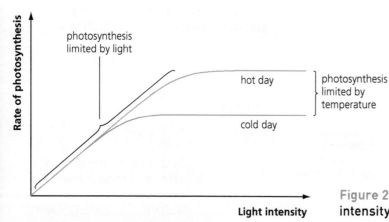

photosynthesis
limited by light

Rate of photosynthesis

hot day

photosynthesis
limited by
temperature

cold day

Light intensity

Figure 2.4 **The effect of increasing light intensity on the rate of photosynthesis**

Exam tip

This experiment could also be repeated with the addition of sodium hydrogen carbonate, which would provide a carbon source for the plant. More photosynthesis would then occur and so more bubbles of oxygen would be seen.

Required practical 2

Investigate photosynthesis, showing the production of starch and the requirements of light and chlorophyll

Equipment:
- A plant with at least two leaves
- Silver foil
- Bunsen burner
- Boiling tube
- Forceps
- Beaker
- Alcohol
- Spotting tile
- Iodine solution

Method:
- One leaf of the plant was covered with a piece of silver foil. The other leaf was positioned so that sunlight fell on it. The plant was then left for two days.
- After two days, both leaves were removed from the plant and heated in boiling water for five minutes.
- Forceps were used to remove both leaves, which were then placed in a beaker of warm alcohol.
- This warm alcohol removed all chlorophyll, which turned the leaf pale.
- Both leaves were then placed on a spotting tile and covered in iodine solution to test for starch.

Results:
- The leaf in the sunlight turned from brown to blue/black with the addition of the iodine solution.
- This showed that starch was present and so the leaf had photosynthesised (the glucose this produced would have been stored as starch).
- The leaf covered in silver foil remained brown with the addition of the iodine solution.
- This showed that starch was not present and therefore the leaf had not photosynthesised (no glucose was produced to be stored as starch).

Now test yourself

TESTED

1 State the word equation for photosynthesis.
2 State the balanced chemical symbol equation for photosynthesis.
3 Besides plants, what other group of organisms completes photosynthesis?
4 Why is photosynthesis an endothermic reaction?
5 In which type of cell does most photosynthesis in leaves occur?
6 What four factors affect the rate of photosynthesis?
7 How do farmers reduce limiting factors for photosynthesis?
8 Define the term limiting factor.
9 State the equipment that you would use to investigate how light intensity affects the rate of photosynthesis.
10 Where are photosynthetic organisms usually found in a food chain?

Answers on p. 124–5

Exam tip

A similar experiment can be done using a leaf from a variegated plant. This type of plant has been bred to have both green and white parts on its leaves to make them look more attractive. The white parts have no chlorophyll so do not photosynthesise. They would not produce glucose or turn it into starch. Therefore, the iodine placed on the white areas would remain brown.

Human nutrition

Balanced diets

A well-balanced diet is a diet that has the correct amounts of all food groups. This can be shown in a food pyramid, as in Figure 2.5. The pyramid shows that:

- A large part of your diet should be made up of carbohydrates like bread, potatoes and rice.
- Another large part of your diet should be fruits and vegetables. These contain natural sugars, healthy vitamins and dietary fibre.
- Only a small part of your diet should be lipids (fats and oils). These are found in dairy products including milk and cheese. Dairy products are also high in protein but have high cholesterol levels, which is why they should be eaten in smaller quantities.

Figure 2.5 A balanced diet represented as a food pyramid.

We don't need the same amount of each food group throughout our lives. In fact, the amount of food we need can change significantly with our activity level, age and condition (for example, pregnancy). People like athletes, or those with more physical jobs, need more food than people who are less active. Similarly, teenagers are growing quickly so require more food, and pregnant women must ensure they have enough food for them and their developing baby.

Table 2.1 provides a summary of the major food groups.

Table 2.1 **The key food groups explained**

Food group	Found in these foods	Function in the body
Carbohydrate	Bread, pasta, rice and potatoes	A source of energy
Protein	Meat, eggs, cheese, fish, nuts and seeds	Growth and repair
Lipids (fats and oils)	Oils, margarine, butter, cheese, cream	A source of energy
Vitamin A	Cod liver oil, liver, ghee, sweet potato, carrot, broccoli	Growth and development, supports the immune system and good vision
Vitamin C	Citrus fruits, broccoli, Brussels sprouts	Repair, making enzymes and supporting the immune system; insufficient vitamin C can lead to scurvy
Vitamin D	Produced by your skin when exposed to sunlight; also found in fish, eggs, liver and mushrooms	Growth of bones; insufficient vitamin D can lead to rickets
Calcium	Milk, cheese, butter, broccoli, cabbage	Growth of bones and teeth, regulates heartbeat and helps blood clot
Iron	Meat, beans, nuts, dried fruit	Healthy blood; insufficient iron in your diet can lead to anaemia
Water	Water, fresh fruit and vegetables	All cells of your body require water to function properly
Dietary fibre	Foods that come from plants like fresh fruit and vegetables	Provides something solid for your digestive system to push along

Revision activity

Draw out Table 2.1 with only the headings and food groups. Try to fill the rest of the table in from memory to help you revise.

The digestive system

Your **digestive system** is about 9 metres long and runs from your mouth to your anus. It breaks down the large, **insoluble** bits of food that you eat into smaller, **soluble** pieces that can be absorbed into your blood. They are then transported around your body to the cells that need them.

Functions of the parts of the digestive system

The digestive system is also known as the **alimentary canal**. The functions of the parts of your digestive system are found in Table 2.2.

Table 2.2 **The parts of the digestive system and their functions**

Component	Function
Salivary glands	Glands in your mouth that produce saliva. Saliva lubricates food as it passes along your oesophagus. It also contains a carbohydrase enzyme called amylase, which begins the breakdown of starch into sugars.
Oesophagus	Short tube that connects the mouth and stomach.
Stomach	Small bag about the size of your fist. It has ridges that allow it to increase in size when you eat food. The food is mixed with stomach acid to kill any **pathogens**. Stomach acid does not break down food. Protease enzymes are mixed with food in the stomach to begin the breakdown of proteins.
Liver	Food does not pass through the liver, but it produces **bile** to break down fats into smaller pieces. This is called emulsification. Breaking fats into smaller pieces increases the surface area to allow lipase enzymes to work more effectively.
Gall bladder	Food does not pass through the gall bladder, but it stores bile before the bile is released into the small intestine.
Pancreas	Food does not pass through the pancreas, but it produces carbohydrase, protease and lipase enzymes, which it releases into the small intestine.
Small intestine	Digested food is absorbed into the blood here. The small intestine is about 7 metres long. The first 30 cm or so is called the duodenum. This is the major site of **breakdown** of food by enzymes. The jejunum is the middle section of the small intestine. Here, most digested food molecules are absorbed into the blood. The ileum is the final 2 to 4 metres of your small intestine. It is the major site of **absorption** of broken down food by **villi**. The surface of the small intestine has millions of these tiny, finger-like projections. They increase the surface area of the small intestine to allow more food to be absorbed into the blood. Food is pushed through your small intestine by **peristalsis**. This is the rhythmical contraction and relaxation of the muscles in the lining of the small intestine. This movement forces lumps of food along it like you might squeeze toothpaste out of a tube. →

Digestive system: The organs and glands in your body that are responsible for breaking down food into smaller, soluble pieces that can be absorbed into your blood.

Insoluble: Things that cannot be dissolved.

Soluble: Things that can be dissolved.

Alimentary canal: Another name for the digestive system.

Pathogen: A disease-causing micro-organism (bacterium, fungus, protist or virus).

Bile: A green/yellow-coloured liquid produced by your liver, stored by your gall bladder and released into your small intestine to break down fats.

Villi (singular: villus): Tiny, finger-like projections that poke into your digestive system in your small intestine to increase the surface area and so absorb more digested food.

Peristalsis: The rhythmical contraction of muscle behind food in the digestive system to push it along.

Component	Function
Large intestine (colon and rectum)	All that is left of your food when it leaves the small intestine is water and fibre that you cannot digest. The colon is the first part of the large intestine, which absorbs water, leaving fibre, which forms your solid waste (faeces).
	The rectum is the final 10 cm of your large intestine. Here faeces is stored until it is removed when you defecate.
Anus	This opening controls when you release faeces when you go to the toilet.

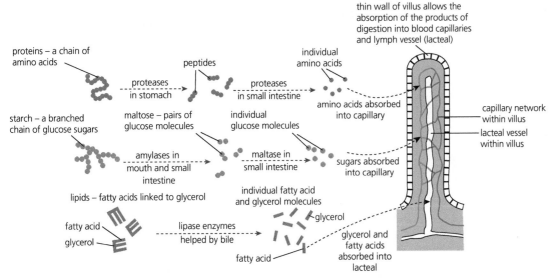

Figure 2.6 **Villi are small, hair-like structures in your small intestine. They increase the surface area over which molecules of digested food can be absorbed.**

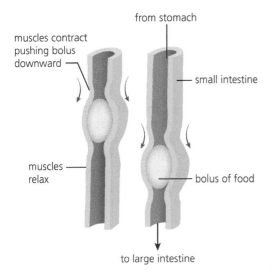

Figure 2.7 **The rhythmical contraction and relaxation of the muscles that line much of the digestive system is called peristalsis**

Revision activity

Draw out a rough sketch of the human body and include the components from Table 2.2. Label each part and write down its function, with arrows showing the movement of both food and other digestive substances.

Exam tip

To help you remember what peristalsis is, picture yourself pushing a tennis ball through a pair of tights. The tennis ball represents a bolus of food and the tights represent your digestive system.

Digestive enzymes

REVISED

There are three types of digestive enzyme. The molecules of food they break down are called substrates (see pages 10–11). The three types of enzyme are summarised in Table 2.3 on the following page.

Table 2.3 The enzymes, substrates and products of the digestive system

Enzyme	Substrate	Product	Location
Carbohydrase	Carbohydrates	Sugars	Mouth, pancreas and small intestine
Protease	Proteins	Amino acids	Stomach, pancreas and small intestine
Lipase	Fats and oils (lipids)	Fatty acids and glycerol	Pancreas and small intestine

Some digestive enzymes have specific names. The carbohydrase enzyme found in your saliva and small intestine is called amylase. This enzyme breaks down the carbohydrate starch into maltose. Another carbohydrase enzyme called maltase is produced by your pancreas. This breaks down the sugar maltose into glucose in the small intestine.

Revision activity

Draw out Table 2.3 with only the headings. Try to fill in the rest of the table from memory.

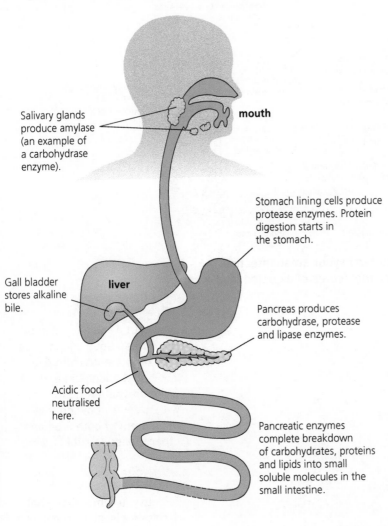

Salivary glands produce amylase (an example of a carbohydrase enzyme).

mouth

Stomach lining cells produce protease enzymes. Protein digestion starts in the stomach.

Gall bladder stores alkaline bile.

liver

Pancreas produces carbohydrase, protease and lipase enzymes.

Acidic food neutralised here.

Pancreatic enzymes complete breakdown of carbohydrates, proteins and lipids into small soluble molecules in the small intestine.

Exam tip

Students often mix up the locations where digestive enzymes are produced (see Table 2.3). It is important that you can describe what the different enzymes do and where they do it, as well as being able to state simple word equations for the reactions of these breakdown enzymes.

Figure 2.8 Digestive enzymes control reactions that take place in the digestive system. No enzymes are made or used in the oesophagus, liver (bile is not an enzyme), gall bladder, large intestine or anus.

Bile

REVISED

Bile is not an enzyme because it does not break down lipids into fatty acids and glycerol like lipase enzymes do. Bile is actually an emulsifier that breaks down large globules of fat into smaller ones. This increases the surface area for lipase enzymes to work on and speeds up their digestion.

Bile is also an alkaline substance that neutralises any excess stomach acid at the beginning of the small intestine. This provides the enzymes in the small intestine with their optimum pH.

Typical mistake

Students often think that bile is an enzyme, but it is not. It is actually an emulsifier – a chemical substance that emulsifies (breaks up) fats.

Required practical 3

Investigate the energy content in a food sample

Equipment:
- Measuring cylinder
- Boiling tube
- Thermometer
- Boss
- Clamp
- Retort stand
- Forceps
- Food samples

Method:
- 5 cm³ of water was placed into a boiling tube.
- A thermometer was used to measure the temperature.
- A boss, clamp and retort stand were used to hold the tube.
- Forceps were used to hold the food sample below the boiling tube.
- The food sample was lit and kept under the tube until it went out.
- The temperature of the water was re-measured and the increase was calculated.
- The experiment was repeated with identical masses of other foods, and the different results were compared.

Results:
- Foods that contained more energy increased the temperature of the water more than foods with lower energy contents.

Now test yourself

TESTED ☐

11 What food groups are present in a balanced diet?
12 What are lipids?
13 Why do pregnant women require more food?
14 Which foods are high in carbohydrate?
15 Why is calcium important in your diet?
16 What does the pancreas do as part of your digestive system?
17 What are the duodenum and ileum part of?
18 Describe the process of peristalsis.
19 What are the three main enzyme types in the human digestive system?
20 Where is bile produced and stored?

Answers on p. 125

Exam tip

The energy values given in the nutritional information on food packaging have been calculated using this technique. Usually an automated machine called a calorimeter is used to burn the food and calculate the energy present. One food calorie is the energy needed to increase 1 cm³ of water by 1 °C.

Respiration

Aerobic respiration

REVISED ☐

Energy is released from glucose during **aerobic** respiration. This reaction is essential for all living organisms. It occurs continuously in specially adapted cell components called mitochondria, which are found in the cytoplasm.

Respiration releases energy, which is used by the cell to convert a chemical substance called ADP to **ATP**. ATP has more energy stored within its chemical bonds than ADP. When the cell needs energy, ATP is converted back to ADP and the extra energy stored in the chemical bonds is released. ATP is often called the 'energy currency' of the cell.

The energy released when ATP is converted back to ADP is used to complete the eight life processes (see page 1). The word equation for aerobic respiration is:

Aerobic: In the presence of oxygen.

ATP (adenosine triphosphate): A molecule found in all cells in which energy from respiration is stored.

$$\text{glucose + oxygen} \xrightarrow{\text{energy out}} \text{carbon dioxide + water}$$

The balanced chemical symbol equation for respiration is:

$$C_6H_{12}O_6 + 6O_2 \xrightarrow{\text{energy out}} 6CO_2 + 6H_2O$$

Respiration is an exothermic reaction so it transfers energy to its surroundings. Unlike photosynthesis, respiration occurs at all times not just during the day.

Conversion of energy in respiration (and photosynthesis)

The equations for photosynthesis and respiration are the opposite of each other: the reactants in respiration are the same as the products in photosynthesis and vice versa. They are also opposite in terms of their energy use. Photosynthesis is endothermic and so takes energy from its surroundings (light) and stores this in glucose. Respiration is exothermic and so releases energy from glucose.

Photosynthesis and respiration work together to support almost all life on our planet. The Sun's energy drives nearly all food chains as its light energy is converted by plants and algae into a chemical store of energy (glucose) during photosynthesis. The energy for life processes is then transferred from glucose during respiration.

This energy has two main uses. It is converted into:
1 heat energy to keep homoeothermic (warm-blooded) animals warm
2 chemical stores of energy that are used in reactions to build larger molecules and drive processes like movement.

Anaerobic respiration in animals

REVISED

Anaerobic respiration occurs in animals when there is not enough oxygen for aerobic respiration. An example might be when you have been exercising vigorously. This means you are unable to breathe quickly and deeply enough to supply your cells with sufficient oxygen. This is the equation for anaerobic respiration in animals:

$$\text{glucose} \xrightarrow{\text{energy out (only 5\%)}} \text{lactic acid}$$

When your cells respire anaerobically, there is insufficient oxygen to make carbon dioxide and water, so a product called lactic acid is made instead. Many scientists think that lactic acid build-up in muscles causes cramp. During long periods of vigorous exercise, your muscles become fatigued and could stop contracting efficiently.

Crucially, only 5% of the energy provided by aerobic respiration is released. The remaining 95% is stored within the lactic acid.

When you have finished exercising, your breathing rate and volume (depth) do not return to normal immediately. They remain high until your body has absorbed enough oxygen to remove the lactic acid that has built up. This is called paying your oxygen debt. This releases the remaining 95% of the energy that was originally stored in the glucose.

Lactic acid builds up in cells that are respiring anaerobically. This lactic acid diffuses into the bloodstream from a higher concentration in the cells to a lower concentration in the blood. It is then transported to the liver where it diffuses from a higher concentration in the blood to a lower concentration in the liver. Once in the liver, it is then converted back to glucose.

> **Exam tip**
>
> When writing the word and symbol equations, the two reactants and the two products can be put either way around. However, you must not mix up reactants with products as this would be incorrect.

> **Exam tip**
>
> When describing cellular respiration, remember to say that it is an exothermic reaction that is continuously occurring in all cells.

> **Exothermic reaction:** A reaction in which energy is transferred to the surroundings. This is the opposite of an endothermic reaction.
>
> **Homoeothermic:** Warm-blooded, like birds and mammals.

> **Anaerobic:** In the absence of oxygen.
>
> **Product:** A substance produced in a reaction.
>
> **Oxygen debt:** A temporary shortage of oxygen in respiring tissues and organs.

> **Exam tip**
>
> When comparing aerobic and anaerobic respiration, you should be specific about the relative amounts of energy that are released.

Anaerobic respiration in plants and micro-organisms

Some micro-organisms such as yeast respire anaerobically:

$$\text{glucose} \xrightarrow{\text{energy out}} \text{ethanol} + \text{carbon dioxide}$$

This reaction in yeast is called fermentation. It is an economically important reaction because we use it to make bread and beer. The ethanol produced is commonly called alcohol, and it is present in concentrations of around 4% in beer, 12% in wine and 40% in spirits.

> **Fermentation**: The chemical breakdown of glucose into ethanol and carbon dioxide by respiring micro-organisms such as yeast.

Required practical 4

Investigate the evolution of carbon dioxide from respiring seeds or other suitable living organisms

Equipment:
- Boiling tubes
- Cotton wool
- Soda lime
- Bungs
- Screw clips
- Capillary tubes
- Invertebrates

Method:
- The equipment was set up as shown in Figure 2.9.
- The distances from the rubber bungs to the water drops in each capillary were measured.
- The experiment was left for one hour.
- The distances to the bubbles were re-measured and the distances they had moved were recorded.

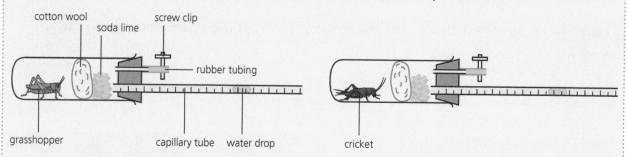

cotton wool   soda lime   screw clip   rubber tubing

grasshopper   capillary tube   water drop   cricket

Figure 2.9 The equipment used to investigate the production of carbon dioxide from living organisms

Results:
- Like all other animals and plants, both the grasshopper and the cricket produced carbon dioxide when they respired.
- This was absorbed by the soda lime as it was produced, which caused a partial vacuum.
- The bubble of water therefore moved towards the bung.
- The greater the respiration of the invertebrates, the further the bubble moved.

Please note: When using animals in experiments it is very important that they are properly looked after. If you carry out this experiment, you should ensure it is completed before the oxygen in the boiling tubes runs out.

Required practical 5

Investigate the evolution of heat from respiring seeds or other suitable living organisms

Equipment:
- Two flasks
- Two thermometers
- Germinating seeds

- Boiled (so non-germinating) seeds
- Cotton wool

Method:
- The equipment was set up as shown in Figure 2.10.
- The starting temperature was measured and the experiment was left for one hour.
- The temperature was re-measured and the difference was calculated.

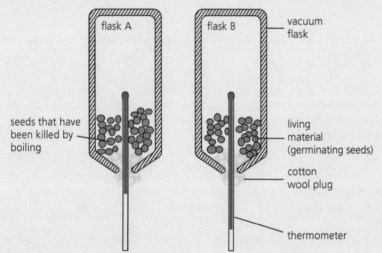

Figure 2.10 The equipment used to show that heat is given off by living material during respiration

Results:
- The germinating seeds respired.
- The temperature in their flask increased as a result.
- The boiled (non-germinating) seeds did not respire and so their temperature remained the same.

Now test yourself

TESTED ☐

21 State the word equation for aerobic respiration.
22 State the balanced chemical symbol equation for aerobic respiration.
23 Why is respiration an exothermic reaction?
24 When do plants complete respiration and photosynthesis?
25 What are the two main uses of the energy produced in respiration in warm-blooded animals?
26 What does anaerobic mean?
27 State the word equation for anaerobic respiration in animals.
28 What is lactic acid converted into?
29 What is anaerobic respiration in yeast also known as?
30 State the word equation for anaerobic respiration in yeast.

Answers on p. 125

Revision activity

Draw a mind map of all the content in this chapter. You could use the topic headings: nutrition in flowering plants, human nutrition, and respiration.

Summary

- The word equation for photosynthesis is:

$$\text{carbon dioxide + water} \xrightarrow{\text{light in}} \text{glucose + oxygen}$$

- The balanced chemical symbol equation for photosynthesis is:

$$6CO_2 + 6H_2O \xrightarrow{\text{light in}} C_6H_{12}O_6 + 6O_2$$

- Photosynthesis is an endothermic reaction in which light energy is transferred to the chloroplasts in plant cells.
- The rate of photosynthesis is affected by temperature, light intensity, carbon dioxide concentration and the amount of chlorophyll.
- These limiting factors can interact. They are important in the economics of obtaining optimum conditions in greenhouses for growing food.
- Glucose produced in photosynthesis is used for respiration, converted into sucrose or starch for storage, and used to produce fats or oils for storage, cellulose for growth and amino acids for protein synthesis.
- A well-balanced diet has the correct amounts of all food groups. Carbohydrates provide an energy source. Proteins are required for growth and repair. Lipids (fats and oils) are another source of energy. Fibre provides solidity for peristalsis. Vitamins and minerals are required in smaller quantities to keep you healthy.
- The digestive system is an organ system that breaks down large lumps of insoluble food into smaller, soluble molecules that are absorbed into the blood. Digested food is absorbed into your blood in your small intestine. Water is absorbed into your blood in your large intestine.
- Enzymes are proteins that speed up reactions including digestion of food.
- Carbohydrase enzymes are made in your mouth, pancreas and small intestine. They break down larger carbohydrates into sugars.
- Protease enzymes are produced in your stomach, pancreas and small intestine. They break down proteins into amino acids.
- Lipase enzymes are made in your pancreas and small intestine. They break down lipids into fatty acids and glycerol.
- Respiration can occur with oxygen (aerobic) or without (anaerobic).
- The energy transferred by respiration supplies all the energy needed by living organisms. It is needed for chemical reactions to build larger molecules, movement and keeping warm (in warm-blooded animals).
- The word equation for aerobic respiration is:

$$\text{glucose + oxygen} \xrightarrow{\text{energy out}} \text{carbon dioxide + water}$$

- The balanced chemical symbol equation for aerobic respiration is:

$$C_6H_{12}O_6 + 6O_2 \xrightarrow{\text{energy out}} 6CO_2 + 6H_2O$$

- During exercise, the heart and breathing rates increase, as does the depth of breathing. This supplies respiring cells with more oxygen. Without enough oxygen, anaerobic respiration occurs.
- The equation for anaerobic respiration in animals is:

$$\text{glucose} \xrightarrow{\text{energy out (only 5\%)}} \text{lactic acid}$$

- Only around 5% of the energy is released in anaerobic respiration. The rest remains within lactic acid. When an oxygen debt has been paid, the rest of the energy is released.
- Lactic acid is transported to the liver where it is converted back to glucose. The oxygen debt is the amount of extra oxygen needed to react with and remove the lactic acid.
- The equation for anaerobic respiration in plant and yeast cells is:

$$\text{glucose} \xrightarrow{\text{energy out}} \text{ethanol + carbon dioxide}$$

- Anaerobic respiration in yeast is called fermentation. This is economically important in the making of bread and alcoholic drinks.

Exam practice

1 Photosynthesis occurs in the green parts of all plants.
  (a) In which of the following cell components does photosynthesis occur?   [1]
       A   Vacuole
       B   Ribosomes
       C   Mitochondria
       D   Chloroplasts
  (b) Explain why photosynthesis is important for all life on Earth, and not just plants and algae.   [2]
  (c) State the word equation for photosynthesis.   [2]
  (d) Describe a procedure you would use to investigate the effects of light intensity on
      photosynthesis.   [4]
  (e) Describe the uses of glucose produced by photosynthesis.   [6]
2 Respiration releases energy for cells to complete life processes.
  (a) What are the product(s) of anaerobic respiration (fermentation) in micro-organisms?   [1]
       A   Lactic acid
       B   Ethanol and carbon dioxide
       C   Water and carbon dioxide
       D   Glucose and oxygen
  (b) Define the term oxygen debt.   [1]
  (c) Compare and contrast the reactions of aerobic respiration and photosynthesis.   [6]
3 Your digestive system is responsible for breaking down food, which is absorbed into your blood.
  Enzymes have an important role in this process.

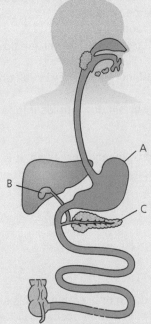

  (a) The diagram shows a human digestive system. Name the parts A to C.   [3]
  (b) Name the **two** products when lipase enzymes break down lipids.   [2]
  (c) Describe how the lock and key theory is used to explain how enzymes break down substrates. [3]
  (d) Explain how extremes of pH or temperature might affect an enzyme's function.   [3]

Answers and quick quizzes online   ONLINE

3 Movement of substances in living organisms

Gas exchange in flowering plants

Gas exchange in leaves

Gases move into and out of leaves by diffusion. Diffusion is the net movement of particles from an area of higher to lower concentration. This process happens naturally and does not require energy.

Plants complete two processes that involve the exchange of gases:
- respiration – this occurs at all times (both day and night)
- photosynthesis – this only occurs during the day.

This means that plants, like animals, are continually taking in oxygen and producing carbon dioxide. The rate at which this process happens is fairly constant.

The equation for aerobic respiration is:

$$C_6H_{12}O_6 + 6O_2 \xrightarrow{\text{energy out}} 6CO_2 + 6H_2O$$

Photosynthesis requires light energy. In this process, plants take in carbon dioxide and produce oxygen.

The equation for photosynthesis is:

$$6CO_2 + 6H_2O \xrightarrow{\text{light in}} C_6H_{12}O_6 + 6O_2$$

Unlike respiration, plants do not photosynthesise at the same rate throughout the day. They photosynthesise more at higher light intensities (which is usually in the middle of the day).

Gas exchange in plants is summarised in Table 3.1.

Table 3.1 **The overall movement of gases into plants at various times of the day**

Time of day	Processes occurring	Movement of gases
Dawn	Respiration	Oxygen in and carbon dioxide out
	Some photosynthesis	A little carbon dioxide in and a little oxygen out
Midday	Respiration	Oxygen in and carbon dioxide out
	More photosynthesis	Carbon dioxide in and oxygen out
Dusk	Respiration	Oxygen in and carbon dioxide out
	Some photosynthesis	A little carbon dioxide in and a little oxygen out
Night	Respiration	Oxygen in and carbon dioxide out

Revision activity

Draw out Table 3.1 with only the headings completed. Try to fill the rest of the table in from memory to help you remember the processes involved.

You can see that more photosynthesis occurs as the Sun rises and the light intensity strengthens. As a result, there is a point each morning when the rate of photosynthesis increases to match exactly the rate of respiration, before the rate of photosynthesis then increases further in the middle of the day. The same occurs in the evening as the light intensity weakens, and the rate of photosynthesis then reduces to match the rate of respiration again. The time at which these two processes occur at equal rates is called the compensation point.

> **Compensation point:** The point at which the rates of photosynthesis and respiration are equal in a plant (often early to mid-morning and mid- to late afternoon).

Structure of the leaf

REVISED

You should have learned about the structure of the leaf in Chapter 2 (see page 22). The structure of the leaf is important in maximising the efficiency of photosynthesis. For example, palisade mesophyll cells are found immediately below the epidermis. These cells contain many chloroplasts to maximise the rate of photosynthesis, and so are ideally positioned near the top of the leaf, where the light is strongest. Below this layer are the spongy mesophyll cells, which have fewer chloroplasts but have air spaces between them to allow gases to diffuse more easily.

Stomata in detail

Stomata are tiny pores found in low numbers on the tops of leaves and in much greater numbers on their undersides. These pores open and close to allow gases to diffuse into and out of the leaf. Each stoma is surrounded by two guard cells that open and close it to regulate the exchange of gases with the air.

During the day, many stomata open to allow gas exchange. The guard cells achieve this by absorbing water through osmosis and becoming turgid. The inner walls of the stomata are rigid, so they are pulled apart when the water is absorbed and the guard cells inflate. At night the reverse happens as water is lost from the guard cells and they become flaccid, causing the pore to close.

> **Turgid:** Swollen or inflated.
>
> **Flaccid:** Drooping or deflated.

The opening of stomata during the day increases the rate of transpiration. This means more water is transported to the leaves in xylem vessels, which allows an increased rate of photosynthesis.

Diffusion and osmosis within the leaf

Water is transported to plant leaves in xylem vessels. This water moves by osmosis from a higher concentration near the xylem vessels to a lower concentration in the cells of the leaf. The cells within the leaf are using water during photosynthesis and so have a permanently lower concentration. Thus, while the plant is photosynthesising, water will always move from xylem vessels to the palisade mesophyll cells.

Carbon dioxide and oxygen diffuse into and out of the leaf via the stomata. Once in the leaf, they diffuse into the air spaces between the spongy mesophyll cells and then move into the cells themselves. The gases naturally diffuse from a higher to a lower concentration into the surrounding palisade mesophyll cells to be used for photosynthesis or respiration.

Required practical 1

Investigate the effect of light on net gas exchange from a leaf, using hydrogencarbonate indicator

Equipment:
- Three boiling tubes and bungs
- Hydrogencarbonate indicator
- Two leaves

Method:
- The equipment was set up as shown in Figure 3.1.
- At the beginning of the experiment, the hydrogencarbonate indicator was orange in all three tubes because air had been bubbled through it.
- The carbon dioxide level was in balance with that in the surrounding air.
- Tube A was placed in bright light and tube B in the dark for half an hour.
- The colour of the tubes was compared (as shown in Figure 3.1).

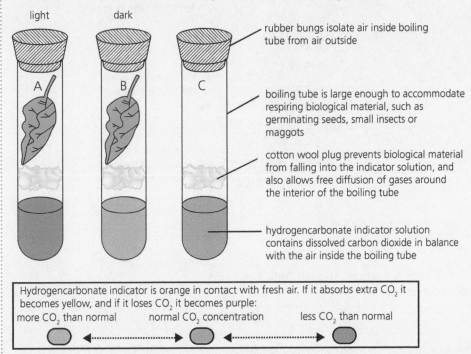

light   dark

rubber bungs isolate air inside boiling tube from air outside

boiling tube is large enough to accommodate respiring biological material, such as germinating seeds, small insects or maggots

cotton wool plug prevents biological material from falling into the indicator solution, and also allows free diffusion of gases around the interior of the boiling tube

hydrogencarbonate indicator solution contains dissolved carbon dioxide in balance with the air inside the boiling tube

Hydrogencarbonate indicator is orange in contact with fresh air. If it absorbs extra CO_2 it becomes yellow, and if it loses CO_2 it becomes purple:
more CO_2 than normal   normal CO_2 concentration   less CO_2 than normal

Figure 3.1 The equipment used to investigate how light affects gas exchange in a leaf

Results:
- The hydrogencarbonate indicator in tube A turned purple, indicating that less carbon dioxide than normal was present. This is because the plant had photosynthesised and so used carbon dioxide.
- Tube B turned yellow, indicating that more carbon dioxide than normal was present. This is because the leaf had only been respiring, not photosynthesising, and so produced carbon dioxide.

Now test yourself

TESTED

1 Which two chemical reactions do all plants complete during the day?
2 Which process results in plants releasing carbon dioxide?
3 State the word equation for photosynthesis.
4 State the chemical symbol equation for aerobic respiration.
5 How do guard cells open?
6 What does turgid mean?
7 What is the compensation point?
8 At what times of the day do the compensation points of plants usually occur?
9 In an experiment investigating the effect of light on gas exchange in a leaf, hydrogencarbonate indicator is orange when carbon dioxide levels are equal with those in the air. What colour would it be at the compensation point of the plant?

Answers on p. 125

Gas exchange in humans

The respiratory system

Your **respiratory system** is responsible for the exchange of gases in your body. It is shown in Figure 3.2.

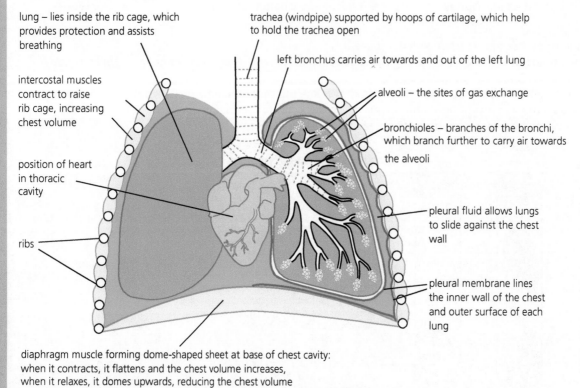

lung – lies inside the rib cage, which provides protection and assists breathing

intercostal muscles contract to raise rib cage, increasing chest volume

position of heart in thoracic cavity

ribs

trachea (windpipe) supported by hoops of cartilage, which help to hold the trachea open

left bronchus carries air towards and out of the left lung

alveoli – the sites of gas exchange

bronchioles – branches of the bronchi, which branch further to carry air towards the alveoli

pleural fluid allows lungs to slide against the chest wall

pleural membrane lines the inner wall of the chest and outer surface of each lung

diaphragm muscle forming dome-shaped sheet at base of chest cavity: when it contracts, it flattens and the chest volume increases, when it relaxes, it domes upwards, reducing the chest volume

Figure 3.2 The parts of your respiratory system. The left lung has been cut away to show the internal structures.

Typical mistake

Some students think that your respiratory system is responsible for respiration in your body. However, respiration is the release of energy from glucose and this is something all of your cells do or else they would die. Your respiratory system, together with your circulatory system, is responsible for providing oxygen to your cells so they can carry out aerobic respiration.

Respiratory system: Your lungs and airways, which are responsible for exchanging carbon dioxide and oxygen.

Abdomen: The part of your body below the thorax, which contains your digestive and reproductive organs.

Your thorax is your chest cavity. This is the part of your body between your neck and your **abdomen**. The parts of your thorax involved in gas exchange and breathing are described in Table 3.2 together with their functions.

Table 3.2 **Some parts of the thorax and their functions**

Part of respiratory system	Structure and function
Trachea	Main airway that leaves the back of your mouth and takes air towards your lungs. Surrounded by rings of cartilage to keep it open at all times.
Bronchus (plural: bronchi)	Your trachea divides into two bronchi, each leading to a lung. These also have cartilage to keep them open.
Bronchiole	Each bronchus divides into thousands of bronchioles. These tubes divide further and further as they take the air deeper into each lung. They are approximately 1 mm in diameter and are not supported by cartilage.
Alveolus (plural: alveoli)	Tiny air sacs found at the ends of bronchioles. There are hundreds of millions in each lung. They have the following adaptations to maximise exchange of gases: ● huge surface area: the total area of your lungs would cover a tennis court ● rich blood supply: there are thousands of tiny blood capillaries in your lungs ● layer of moisture inside to increase rate of diffusion ● lining one cell thick to increase rate of diffusion.
Pleural membranes	Form a double layer around your lungs and hold pleural fluid between them. This allows smooth movement as your lungs increase and decrease in size as you breathe.
Intercostal muscles	Found in between your ribs, they contract and relax during breathing.
Ribs	Curved bones that protect your heart, lungs and other vital organs in your thorax.
Diaphragm	Large, domed-shaped sheet of muscle that separates your lungs from your abdomen; used in breathing.

Revision activity

Make a set of flash cards with a part of the respiratory system on one side of each card and its function on the other. Try to remember what each part does before turning over to see if you are right.

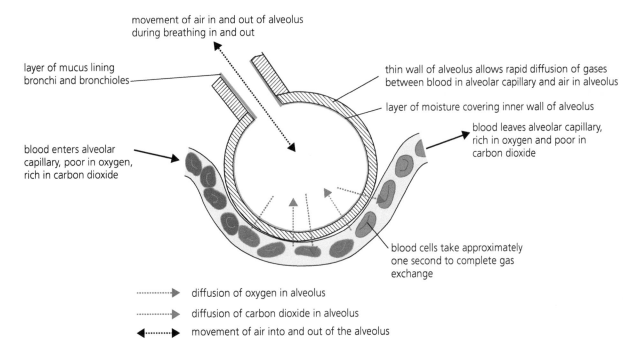

Figure 3.3 **Gas exchange in an alveolus**

Ventilation

Ventilation is the process of breathing, and it involves your diaphragm and intercostal muscles. When breathing normally you do not suck air into your lungs or forcibly expel it. The movement of air into and out of your lungs is normally a passive process. However, if you are exercising and panting to get more oxygen into your lungs quickly, this will cause you to suck air in and blow it out.

Inhalation is the process of breathing in, and it happens as follows:
- Your intercostal muscles contract, forcing your rib cage up and outwards.
- Your diaphragm muscle contracts, which pulls it downwards.
- These movements increase the volume of your chest, which reduces the pressure inside your lungs.
- This causes air to be pulled in.

Exhalation is the process of breathing out, and it happens as follows:
- Your intercostal muscles relax, moving your rib cage down and inwards.
- Your diaphragm muscle relaxes, which causes it to go upwards.
- These movements reduce the volume of your chest, which increases the pressure inside your lungs.
- This causes air to be pushed out.

Table 3.3 shows how the air you breathe in is different from the air you breathe out. The reduction in oxygen content of around 5% is because this is what you use during respiration. The increase in carbon dioxide content of around 100 times is because it is produced during respiration.

Table 3.3 The gases in inhaled and exhaled air. Other gases in small amounts make up the remainder.

Gas	Inhaled air	Exhaled air
Nitrogen	78%	78%
Oxygen	21%	16%
Carbon dioxide	0.004%	4%

> **Ventilation:** The process of breathing in humans, which exchanges carbon dioxide with oxygen.
>
> **Inhalation:** The process of breathing in.
>
> **Exhalation:** The process of breathing out.

> **Revision activity**
>
> Draw two flow diagrams for inhalation and exhalation to help you remember the processes.

> **Revision activity**
>
> Draw out Table 3.3 with only the headings. Try to fill the rest of the table in from memory to help you remember.

Required practical 2

Investigate breathing in humans, including the effect of exercise

Method:
- Using a stopwatch, a student recorded the number of breaths they took in one minute.
- The student completed five minutes of strenuous exercise.
- They then recorded the number of breaths each minute until their breathing rate returned to their previous value.

Results:
- A graph of breathing rate per minute against time after exercise was plotted. The breathing rate increased during exercise, as more oxygen was needed by the respiring cells.
- As the student recovered from the exercise, their breathing rate returned to normal.

Required practical 3

Investigate breathing in humans, including the release of carbon dioxide

Equipment:
- Boiling tube
- Limewater
- Straw

Method:
- A boiling tube was half-filled with limewater.
- A straw was used to gently blow exhaled air into the limewater.
- Any change in colour was recorded.

Results:
- The limewater changed from clear to cloudy.
- This proved that carbon dioxide is present in exhaled air.

Extension: This could also be completed with hydrogencarbonate indictor, which changes from orange to yellow in the presence of carbon dioxide.

Smoking and the respiratory system

Smoking reduces a person's ability to exchange gases in their lungs. This makes it harder for smokers to provide their cells with the oxygen they need for respiration. As a result, smokers are often tired or short of energy, especially when exercising. Heavy smokers often find themselves short of breath.

In some parts of the world, air pollution is so bad that it contributes to problems with people's respiratory systems. However, smoking is still the biggest risk factor for respiratory disease. In the UK, more than 80% of people suffering from respiratory disease are smokers or ex-smokers.

Emphysema

Emphysema is a medical condition where a person's alveoli are damaged, often as a result of smoking. This reduces the surface area in the lungs, which reduces the rate of gas exchange. People with emphysema are often breathless or wheezy, have a chesty cough (see the section on 'Smoker's cough' below) and can have frequent chest infections. This damage to the alveoli cannot be reversed. Stopping smoking is the best way of minimising any further damage.

Chronic obstructive pulmonary disease (COPD) is a general term used to describe emphysema and other lung problems including bronchitis (inflammation of the airways).

Smoker's cough

Smoking involves sucking hot, acidic gases into the respiratory system. These gases damage the cells that line the trachea, bronchi and bronchioles, as well as the alveoli.

Two types of special cell which can be damaged by smoking line your airways. These are:
- Goblet cells, which produce mucus to trap dust, dirt and pathogens.
- Ciliated cells, which have tiny hairs called cilia that move like a Mexican wave to remove the mucus. The mucus is moved by the

Emphysema: A medical condition that makes gas exchange difficult due to damaged alveoli and therefore a reduced surface area in the lungs.

Goblet cells: Cells lining your airways that produce mucus to trap dust, dirt and pathogens.

Ciliated cells: Cells lining your airways that have tiny hairs called cilia.

Cilia: Tiny hairs found on ciliated cells lining your airways, which beat rhythmically to remove mucus.

cilia up your trachea to your throat, where you swallow it into your stomach. Most pathogens are then destroyed by your stomach acid.

Smoking reduces the effectiveness of ciliated cells and results in a smoker's cough.

Coronary heart disease

Communicable: A disease that can be transmitted from one organism to another.

Communicable diseases like the common cold can be transmitted between people. Coronary heart disease is a non-communicable disease. This means it develops, rather than being caught.

Your heart is like all the other organs in your body. Its cells need glucose and oxygen for respiration. These are delivered to your heart through its **coronary arteries**.

Unhealthy lifestyles, such as little exercise, smoking, drinking excessively and a poor diet, can result in a build-up of fat inside coronary arteries. It also reduces the flexibility of their linings. This can slow or stop blood from reaching the heart, which can result in heart attacks.

Coronary heart disease is now one of the major causes of death in the world. A treatment for this disease that used to be common was a **heart bypass** operation. This involved moving a short section of artery from another part of the body and using it to allow blood to flow around the blockage. This is a major form of surgery and so has inevitable risks. More recent treatments are less risky and include **stents** and statins. Stents are small medical meshes that can be expanded to hold open blocked arteries. Statins are drugs prescribed by doctors that reduce **cholesterol**, and so reduce coronary heart disease.

If all other treatments fail, some patients need a heart transplant, which is a serious medical procedure. All transplants need a match between the donor's organ and that of the patient. This means some patients can be left on waiting lists for years.

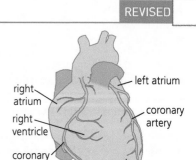

Figure 3.4 **The coronary arteries supply glucose and oxygen to the heart muscle tissue.**

Coronary arteries: Arteries that supply the heart muscle with oxygenated blood.

Heart bypass: A medical procedure in which a section of less important artery (often from the leg) is inserted around a blockage in a coronary artery to allow blood to flow past it.

Stent: A small medical device made from mesh that keeps arteries open.

Cholesterol: An important biological molecule needed to produce cell membranes but which can cause fat build-up if levels are too high in the blood.

Now test yourself

10 What system is responsible for gas exchange in humans?
11 Between which two parts of your body is your thorax located?
12 How is your trachea adapted for its function?
13 Why are the insides of your lungs moist?
14 What happens to your intercostal muscles and diaphragm when you inhale?
15 How is the amount of carbon dioxide different in inhaled and exhaled air?
16 Define the term emphysema.
17 What two cell types line the airways?
18 State the main cause of coronary heart disease.
19 What is the main advantage of using stents rather than bypass operations in treating coronary heart disease?

Answers on p. 125

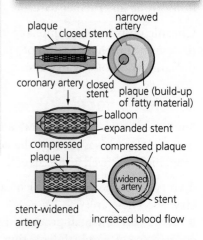

Figure 3.5 **A stent allows blood to flow freely again.**

Transport in living organisms

Diffusion, osmosis and active transport

There are three processes by which cells can absorb the substances (oxygen, carbon dioxide, glucose and water) they need to survive:

1 Diffusion – the net movement of particles from a higher to a lower concentration
2 Osmosis – the net diffusion of water across a partially permeable membrane
3 Active transport – the net movement of substances from a lower to a higher concentration using energy.

These processes allow substances to move into **unicellular** organisms, as well as between the cells of larger, multicellular organisms.

> **Unicellular:** An organism such as a bacterium that is made from one cell only.

When looking at the size of a whole organism, smaller (especially unicellular) organisms have a much larger surface area to volume ratio than larger organisms. The ratio reduces as organisms increase in overall size. This relationship is shown in Figure 3.6.

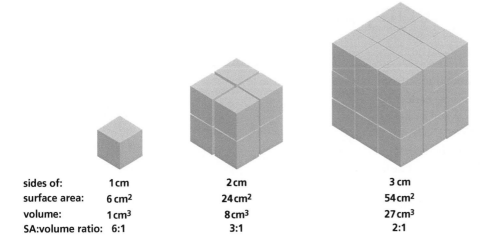

sides of:	1 cm	2 cm	3 cm
surface area:	6 cm²	24 cm²	54 cm²
volume:	1 cm³	8 cm³	27 cm³
SA:volume ratio:	6:1	3:1	2:1

Figure 3.6 **The surface area to volume ratio reduces significantly in larger organisms.**

Unicellular organisms like yeast and photosynthesising protoctists like *Chlorella* have a large surface area to volume ratio. This means that the substances they need can move directly into them. Yeast requires oxygen and a sugar like glucose for growth. These substances move by diffusion into the cells. Yeast also requires water, which moves by osmosis from its surroundings. *Chlorella* needs carbon dioxide for photosynthesis during the day and oxygen for respiration at all times. These gases move by diffusion into its cells. It also needs water, which again enters by osmosis.

Larger organisms have a significantly smaller surface area to volume ratio. They also have more cells within them that are further away from the surface. This means substances have to travel further to reach all cells. Larger organisms have therefore developed transport systems. For example:

● Plants have xylem and phloem vessels to transport water and the sugars made in photosynthesis.

- Mammals have a respiratory system including lungs and a circulatory system.
- Fish have a respiratory system including gills and a circulatory system.

Without these systems, larger organisms cannot exist. Smaller organisms such as insects can use a very simple gas exchange system, which consists of tubes called trachea running between their tissues. These trachea open to the air via small holes called spiracles. Gases diffuse into and out of these holes. This system is less efficient than lungs in mammals and gills in fish, so it is unlikely that insects could grow to a very large size.

Now test yourself

TESTED

20 How is active transport different from diffusion?
21 How does the surface area to volume ratio change when the size of an organism increases?
22 Why don't small organisms require transport systems?

Answers on p. 125

Transport in flowering plants

Xylem

REVISED

Xylem tubes carry water and mineral ions from plant roots to their leaves in a process called transpiration. This process is covered in detail later in this section (see pages 45-6). Xylem tissue is made from dead cells with ends that have eroded away to form hollow tubes. These tubes are strengthened by lignin to allow them to transport water to the leaves for photosynthesis.

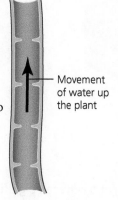

— Movement of water up the plant

Figure 3.7 **A section through a xylem tube**

Xylem: Dead plant cells joined together into long tubes through which water flows during transpiration.

Transpiration: The gradual release of water vapour from leaves to continue the 'pull' of water up to them from the soil.

Phloem and translocation

REVISED

Phloem vessels carry sugars like sucrose that have been dissolved in water. Sugars are made during photosynthesis and are used in a number of ways. They are:

- broken down to release the energy stored in glucose during respiration
- stored as starch or sucrose
- made into amino acids for protein synthesis
- turned into cellulose, fats and oils.

Phloem tubes carry sugars in a process called translocation. Phloem tissues are made from tubes of elongated, living cells. The cells have specialised endings called sieve plates, which allow water to flow more easily through them.

Phloem: Living cells that form tubes to carry the sugars made in photosynthesis to all the cells of a plant.

Translocation: The movement of sugars made in photosynthesis from the leaves of the plant to the rest of its parts.

Xylem and phloem tubes are often found together in structures called vascular bundles. These are the stringy bits you might see in celery, for example.

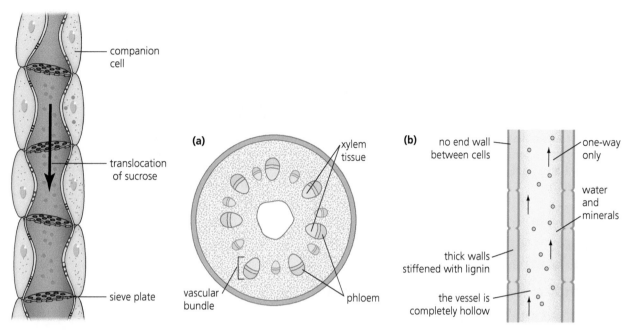

Figure 3.8 **The parts of a phloem tube**

Figure 3.9 **(a) Transverse section of a stem, (b) longitudinal section of xylem vessel from a vascular bundle (the arrows show the direction of water flow)**

Osmosis and root hair cells

REVISED

Root hair cells extend a long way into the soil to massively increase the surface area of the cell. This means it can absorb more water and minerals from the soil. Water is absorbed by osmosis from a higher concentration in the soil to a lower concentration in the root hair cells. Water then moves from the roots through xylem cells due to transpiration, which keeps the concentration in the roots low. Unlike water, mineral ions are absorbed from a lower concentration in the soil to a higher concentration in the plant by active transport, which requires energy.

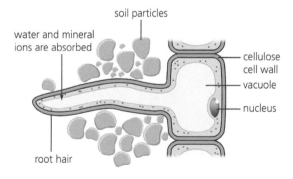

Figure 3.10 **The parts of a root hair cell**

Transpiration

REVISED

Plants do not have a heart so they cannot pump water to their leaves for photosynthesis. Instead it is 'pulled' upwards in a transpiration stream. Water is allowed to evaporate from the stomata, with the guard cells controlling how quickly this happens. As the water evaporates, more is pulled up from the roots to replace it. This process occurs continuously throughout the lives of all plants.

Transporting water by transpiration allows plant cells to remain turgid, and it is vital for photosynthesis and to carry essential minerals from the soil around the plant.

Transpiration rates increase when:
- there is more wind
- the air is drier (less **humid**)
- the temperature is higher
- the light intensity is higher (on sunny days).

Humid: Describes an atmosphere with high levels of water vapour.

Potometer: The equipment used to measure the rate of transpiration in a plant.

Exam tip

You need to be able to explain the effect of changing temperature, humidity, air movement and light intensity on the rate of transpiration.

Typical mistake

Students often confuse the processes of transpiration and translocation. Make sure you are clear about what each one transports and the direction in which it moves.

Required practical 4

Investigate the role of environmental factors in determining the rate of transpiration from a leafy shoot

Equipment:
- Potometer
- Plant shoot
- Water

Method:
- The equipment was set up as shown in Figure 3.11. (This set-up is called a **potometer**.)
- The experiment was left for one hour.
- The distance the water had moved along the capillary tube was measured.
- The experiment was repeated again at a higher temperature, in and out of direct sunlight, in humid and dry conditions and in windy and still conditions.

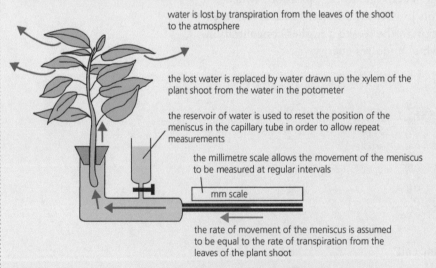

water is lost by transpiration from the leaves of the shoot to the atmosphere

the lost water is replaced by water drawn up the xylem of the plant shoot from the water in the potometer

the reservoir of water is used to reset the position of the meniscus in the capillary tube in order to allow repeat measurements

the millimetre scale allows the movement of the meniscus to be measured at regular intervals

mm scale

the rate of movement of the meniscus is assumed to be equal to the rate of transpiration from the leaves of the plant shoot

Figure 3.11 The equipment used to investigate transpiration

Results:
- The bubble moved further in the same period of time when more transpiration occurred.
- More transpiration occurred at higher temperatures, higher light intensities (in direct sunlight), in dry conditions and in windy conditions.

Transport in humans (1) – blood, structures and functions

Blood

REVISED ☐

You have about 5 litres of blood in your body. It looks red because your blood contains millions of red blood cells. Blood also carries other types of cells and substances, including those absorbed into your blood by your digestive system. These include glucose, fatty acids, glycerol and amino acids.

Red blood cells

There are millions of red blood cells in every drop of your blood. These cells carry oxygen from your lungs to your tissues and organs where it is needed for respiration. Red blood cells have a characteristic **biconcave** shape – this means they have indentations on each side. This shape increases their surface area so they can absorb oxygen from your lungs at a faster rate.

Red blood cells contain a compound called **haemoglobin**. In your lungs, this compound binds with oxygen to form **oxyhaemoglobin**. This turns the colour of your blood from dark red to bright red. When the red blood cells reach your tissues, they release the oxygen and the oxyhaemoglobin turns back to haemoglobin. Red blood cells have no nucleus, which maximises the volume of oxygen they can carry.

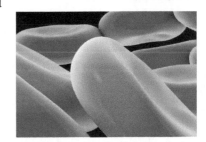

Figure 3.12 **The biconcave shape of red blood cells maximises their surface area to volume ratio and so increases the oxygen they can absorb.**

Biconcave: Describes a shape with a dip that curves inwards on each side.

Haemoglobin: The molecule in red blood cells that can temporarily bind with oxygen to allow it to be carried around the body.

Oxyhaemoglobin: A substance formed when your red blood cells temporarily bind with oxygen.

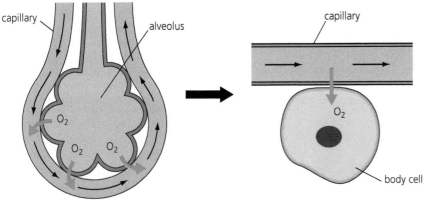

haemoglobin+oxygen → oxyhaemoglobin   oxyhaemoglobin → oxygen+haemoglobin

Figure 3.13 **Haemoglobin in red blood cells transports oxygen as oxyhaemoglobin from the lungs to other organs.**

White blood cells

White blood cells are part of your immune system. They attack pathogens before they can make you ill. You have far fewer white blood cells than red ones. A drop of your blood will only contain tens of thousands of white blood cells compared to millions of red blood cells.

There are two main types of white blood cell:
1 **Phagocytes** – these engulf pathogens by surrounding them and absorbing them. Enzymes inside the phagocytes then destroy the pathogens. Phagocytes are non-specific, so they will attack any pathogen. This process is covered in more detail on the next page.
2 **Lymphocytes** – these make **antibodies**, which join up with the **antigens** found on pathogens. This clumps the pathogens together and makes it easier for them to be destroyed by phagocytes. Antigens are specific to the pathogen they are found on, so both antibodies and lymphocytes are specific to particular pathogens.

Plasma

Blood plasma is a straw-coloured fluid that carries the blood cells, platelets and all other substances around your body. Just over half of your blood is plasma and nearly all of plasma is water. The carbon dioxide produced in respiration dissolves in the plasma and is later removed from the blood in the lungs. Plasma also carries waste urea to your kidneys to be removed in urine, as well as **hormones** made in your glands and heat energy to keep your body temperature as constant as possible.

Platelets and blood clotting

Platelets are cell fragments. There are over 100 000 of them in a drop of blood. Platelets join together when the skin is cut to form a scab. They also release clotting factors that convert a chemical called fibrinogen in your blood into fibrin. Fibrin forms a mesh and sticks platelets to it to form a scab. This prevents blood loss and stops micro-organisms entering the body.

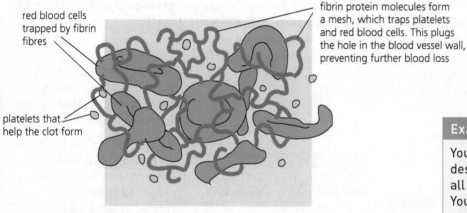

red blood cells trapped by fibrin fibres

fibrin protein molecules form a mesh, which traps platelets and red blood cells. This plugs the hole in the blood vessel wall, preventing further blood loss

platelets that help the clot form

Figure 3.14 **Formation of a blood clot**

> **Phagocyte:** A type of white blood cell that engulfs and destroys pathogens.
>
> **Lymphocyte:** A type of white blood cell that produces antibodies to help clump pathogens together to make them easier to destroy.
>
> **Antibody:** A protein produced by lymphocytes that recognises the antigens of pathogens and helps to clump them together.
>
> **Antigen:** A molecule on the surface of a pathogen that your antibodies recognise as foreign.
>
> **Hormone:** A chemical (produced in a gland in mammals) that moves around an organism to change the function of target cells, tissues or organs.
>
> **Platelets:** Small structures (not cells) in your blood that fuse together to form a scab at the site of a cut.

> **Exam tip**
>
> You should be able to describe the functions of all the blood components. You should also be able to explain how their structure is related to their function.

> **Revision activity**
>
> Try making flash cards of the different blood components. On one side of each card draw a component and on the other side write its functions. For each one, try to remember the functions before turning over to check whether you are right.

The immune system

The first line of defence

The first line of defence stops pathogens from entering your body. These defences are non-specific and are described in Table 3.4.

> **Lysozymes:** Antibacterial enzymes found in your tears and saliva.

Table 3.4 The first line of defence against infection by pathogens

Type	Description
Skin	Provides a barrier that almost completely covers you to prevent attack by pathogens. When your skin is broken with a cut, a scab forms to prevent pathogens entering.
Eyes and mouth	Produce enzymes called lysozymes that attack bacteria by breaking down their cell walls.
Nose	Has hairs and produces mucus to trap any pathogens you breathe in.
Trachea and bronchi	The cells lining your airways have tiny hair-like projections called cilia. In between these ciliated cells are goblet cells that produce mucus. This traps any pathogens that have bypassed the hairs and mucus in your nose. The hairs of the ciliated cells beat in a rhythmical motion to waft the mucus and its trapped pathogens towards your throat. When you clear your throat, you swallow this mucus into your stomach where any pathogens are killed by stomach acid.
Stomach	Your stomach contains hydrochloric acid. This does not break down food but is strong enough to kill many pathogens that enter through your mouth or nose.

The second line of defence

If a pathogen passes your first line of defence, it is attacked by your second line of defence. This is also non-specific, so all pathogens are attacked in the same way. White blood cells called phagocytes attack all pathogens that evade your first line of defence. Their cell membrane flows around the pathogens, engulfing them in a vacuole. Enzymes within their cytoplasm then attack the pathogen cell walls and membranes. This process is called **phagocytosis.**

> **Revision activity**
>
> Draw out Table 3.4 with only the headings and parts filled in. Try to complete the description column from memory to aid your revision.

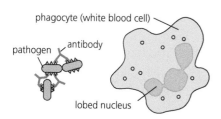 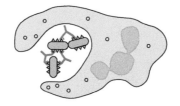 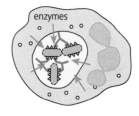

(a) Antibodies cause pathogens to clump together

(b) Phagocyte flows around pathogens to engulf them in a vacuole

(c) Enzymes added to vacuole to break down pathogen cell walls and membranes

Figure 3.15 A white blood cell (phagocyte) engulfing pathogens

The third line of defence

The third line of defence attacks pathogens in a more specific way. Another type of white blood cell called a lymphocyte produces antibodies that specifically match the antigens found on the outside of pathogens. Every pathogen has a different antigen and so every time you are infected by a different pathogen, your lymphocytes need to produce different antibodies. The antibodies cause pathogens to stick together, making it easier for phagocytes to engulf and destroy them.

After exposure to a particular antigen, your lymphocytes form memory lymphocytes. Your memory lymphocytes will 'remember' each pathogen

> **Phagocytosis:** The process of a phagocyte (white blood cell) engulfing and destroying a pathogen.

so that they can produce more antibodies faster if exposed to the same pathogen again. This means you won't catch the same common cold each winter – there are in fact several hundred different common cold viruses.

Your lymphocytes also produce **antitoxins**. These are a special type of protein that can neutralise the toxins produced by some pathogens that make you feel ill.

Antitoxin: A protein produced by your body to neutralise harmful toxins produced by pathogens.

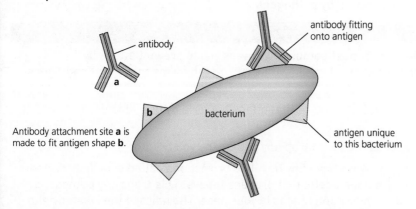

Figure 3.16 How an antibody fits onto the antigen of a pathogen

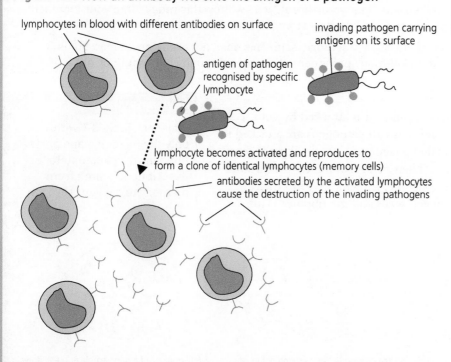

Figure 3.17 A summary of the immune response to an invading pathogen

Vaccination

REVISED

Vaccinations protect us from disease. If a large proportion of a population is vaccinated, it is much less likely that a disease will spread. This is called herd immunity. A vaccination is a small quantity of an inactive form of a pathogen. This is introduced into the body, often by injection. The antigens on the pathogen stimulate lymphocytes to produce antibodies and memory lymphocytes.

Shortly after a vaccination you may feel a little ill. This is your body fighting the disease and is called the initial exposure. However, if you encounter a more severe case of the pathogen later (a secondary exposure), your memory lymphocytes will 'remember' it and produce more antibodies sooner and faster. That means you are much less likely to fall ill.

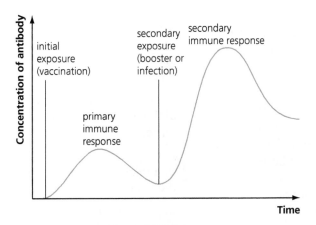

Figure 3.18 This graph shows the rate at which antibodies are produced after the first and second exposures to a pathogen.

Now test yourself

TESTED

32 What are the four main components of blood?
33 Why are red blood cells biconcave in shape?
34 What compound in red blood cells joins with oxygen?
35 How is the structure of red blood cells different from that of the other cells in your body?
36 What is the function of phagocytes?
37 What is the function of lymphocytes?
38 What name is given to the foreign molecules that exist on pathogens and trigger an immune response?
39 What do antitoxins do?
40 What is a vaccine?
41 What is the importance of herd immunity?

Answers on p. 125

Transport in humans (2) – heart and blood circulation

The heart

REVISED

The human heart is an organ made from muscle and nervous tissue. It pumps roughly every second and will beat an amazing 100 000 times a day, or over two billion times in an average lifetime.

Your heartbeat is controlled by your 'natural pacemaker'. This is a small group of nervous tissue cells in the top-right chamber of your heart. They generate an electrical signal that spreads out along your heart's nerve fibres and causes the heart muscle to contract. This contraction pumps blood from your heart. Your pacemaker controls the rate at which your heart beats.

Blood takes about one minute to complete a full circuit of your body. During this time, it will pass through your heart twice. Blood is pumped from your heart to your lungs, then back to your heart and on to the rest of your body, before returning to your heart to start over again. This is called a double circulation.

A human heart has four chambers, two on each side. The top two chambers are called **atria** (the left atrium and right atrium). The bottom two chambers are called **ventricles**. The atria and ventricles are separated by valves.

Blood collects in the atria when the valves are closed. When the heart beats, the blood in the atria is forced into the ventricles. More valves at the ends of the ventricles stop blood being pumped straight into the blood vessels. When the heart beats again, the blood in the ventricles is forced into the blood vessels to begin another journey around the body.

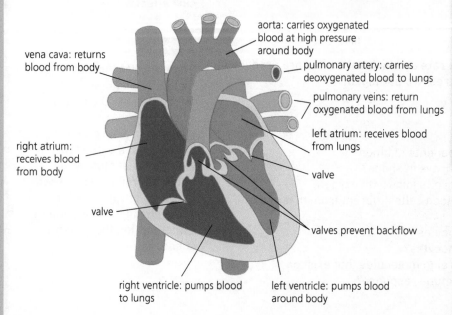

vena cava: returns blood from body

aorta: carries oxygenated blood at high pressure around body

pulmonary artery: carries deoxygenated blood to lungs

pulmonary veins: return oxygenated blood from lungs

left atrium: receives blood from lungs

right atrium: receives blood from body

valve

valve

valves prevent backflow

right ventricle: pumps blood to lungs

left ventricle: pumps blood around body

Figure 3.19 A cross-section of the heart, showing its structure

Blood is pumped from the left atrium into the left ventricle. It is then pumped into an **artery** called the aorta. This divides into smaller arteries and eventually into tiny **capillaries**, which permeate the body's tissues and organs to provide oxygen, glucose and other substances.

The capillaries then collect into small **veins**, which join to become larger, and eventually come together in veins called the venae cavae (singular: vena cava). The venae cavae return the blood to the right atrium. From here it is pumped into the right ventricle and then an artery called the pulmonary artery. This takes blood to the lungs to add oxygen and remove carbon dioxide. The blood then completes its journey by returning to the left atrium in your pulmonary vein.

The muscular lining of your heart is thicker on the left-hand side. This is because the left ventricle pumps blood to all tissues of your body including those in your extremities. The right ventricle only pumps blood to the lungs so it doesn't need such a thick muscular lining.

Coronary heart disease occurs when the major arteries supplying blood to the heart muscle become blocked. This can cause a heart attack. It can be treated by stents or a heart bypass operation (see page 42).

Increasing your heart rate

If you exercise vigorously, your heart rate increases. This ensures more oxygenated blood carrying dissolved glucose is pumped to the cells that need it for respiration (often the muscle cells).

Atrium (plural: atria): An upper chamber of the heart surrounded by a smaller lining of muscle.

Ventricle: A lower chamber of the heart surrounded by a larger lining of muscle.

Artery: A large blood vessel that takes blood away from the heart.

Capillaries: Tiny blood vessels found between arteries and veins. They carry blood into tissues and organs.

Vein: A blood vessel that returns blood to the heart.

Typical mistake

Convention dictates that labelled heart diagrams have the right-hand side on the left and vice versa. You may well have to draw or label a heart diagram in the exam, so make sure you remember the normal convention because it is easy to mix this up.

Revision activity

Write out a flow diagram to show how blood moves around the body. Start and end your diagram in the same place.

Exam tip

Exam questions often ask why the muscle on the left-hand side of the heart is thicker. You should remember that this is because the left-hand part has to pump blood further (to the whole body) than the right (only to the lungs).

Just above the kidneys are the **adrenal glands**, which produce **adrenaline**. Unlike some other hormonal responses, the body responds very quickly to adrenaline. Adrenaline is produced when the body perceives a threat. The response it generates is often called the 'fight or flight' response. The hormone increases the heart rate, providing the muscles with more oxygen and glucose for respiration. This releases more energy in order to fight or flight (run away).

> **Adrenal glands:** Glands found above the kidneys that produce adrenaline.
>
> **Adrenaline:** A hormone produced by your adrenal glands that causes an increase in heart rate ready for a 'fight or flight' response.

The circulatory system

The circulatory system is made up of your heart and all your veins, arteries and capillaries. Its function is to provide all the cells in the body with the substances they need. For example, all cells need glucose and oxygen for respiration and all cells need carbon dioxide to be removed. Some cells will also require different substances like hormones at specific times like puberty. The circulatory system ensures all your cells have all the substances they need.

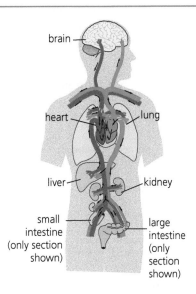

Figure 3.20 **The circulatory system, showing the main organs with which the blood exchanges substances. Bright red is oxygenated blood and dark red is deoxygenated blood.**

Blood vessels

You have between 50 000 and 100 000 miles of blood vessels in your body, which is enough to circle the Earth. There are three main types: arteries, capillaries and veins.

Arteries take blood away from the heart. This blood is under high pressure, so the lining of arteries is thick and muscular. They are also elastic so they can stretch when blood is pumped from the heart. You can feel this surge of blood yourself in places like your wrist where arteries are near the surface. This is your pulse.

The arteries that lead to the lungs are called pulmonary arteries. Those that lead to your kidneys are called renal arteries. The artery that leads to your liver is called the hepatic artery.

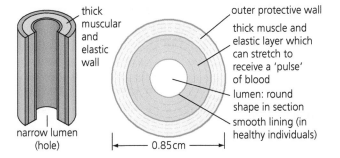

Figure 3.21 **The structure of an artery. Note the thickness of the artery muscle.**

Capillaries carry blood into tissues and organs to provide all the cells with the substances they need. They are very small and thin, and they spread out in tissues and organs like the roots of a tree. Blood plasma passes from these capillaries into tissues, where it is called tissue fluid. It provides cells with glucose and oxygen for respiration. The glucose and oxygen move into the cells by diffusion. Waste products such as carbon dioxide diffuse into the tissue fluid to be carried away in the capillaries.

Veins carry blood back to the heart under much lower pressure. Blood loses pressure during its journey through the capillaries, so the linings of veins do not need to be as muscular as those of arteries. They also have one-way valves (not found in arteries) to keep blood flowing back to the heart. Without these valves, blood would flow much less effectively.

The veins that come from your lungs are called pulmonary veins. Those that come from your kidneys are called renal veins. The vein that comes from your liver is called the hepatic vein.

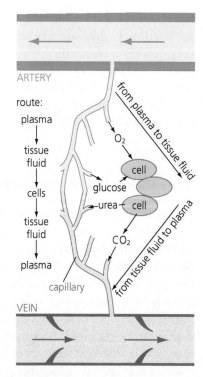

Figure 3.22 Exchange between the blood and tissue cells in a capillary network

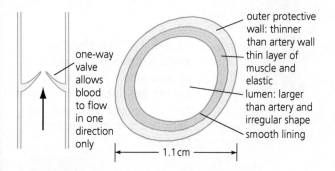

Figure 3.23 The structure of a vein. Note the irregular shape and the thinner muscle and elastic layer of the vein.

Exam tip

You should be able to explain how the structure of blood vessels relates to their function. It is often helpful to link their structures to the pressure of the blood within them.

Revision activity

You could make flash cards of the three blood vessels. Each could have a labelled diagram on one side and how the vessel is adapted for its function on the other. Try to remember the adaptation before turning each card over to check whether you were right.

Now test yourself

TESTED ☐

42 What are the top chambers of your heart called?
43 What stops blood flowing backwards in your heart?
44 Why is the muscular lining of your left ventricle thicker than the right?
45 Why are humans described as having a double circulation?
46 What controls your heartbeat?
47 In what direction do arteries carry blood?
48 How are veins adapted for their function?
49 What effect does adrenaline have on your heart?
50 Define the term capillaries.
51 What term is used for blood vessels that take blood to and from the lungs?

Answers on p. 126

Revision activity

Draw a mind map of all the content in this chapter. You could use the topic headings: gas exchange in flowering plants, gas exchange in humans, transport in living organisms, transport in flowering plants, transport in humans: blood, structures and functions, and transport in humans: heart and blood circulation.

Summary

- Gases move into and out of plant leaves by diffusion.
- Plants only photosynthesise during the day but respire at all times. At night they only release carbon dioxide from respiration. During the day they release both oxygen from photosynthesis and carbon dioxide from respiration.
- Your respiratory system is responsible for gas exchange in your body. It is made up of your airways, trachea, bronchi, bronchioles, and the alveoli in your lungs.
- Smoking reduces the surface area of lungs (causing emphysema, which can result in a characteristic cough) and increases the chance of coronary heart disease.
- All life requires water to survive. Most organisms require oxygen for respiration, which produces carbon dioxide. Photosynthesising organisms need carbon dioxide and water for photosynthesis, which produces glucose and oxygen. All non-photosynthesising organisms need a food source.
- Small organisms have a much larger surface area to volume ratio. This means they do not require the specialised transport systems, such as lungs, found in larger organisms.
- Root hair cells in plants extend far into the soil. This increases the surface area for absorption of water by osmosis and mineral ions by active transport.
- Xylem cells form long tubes that carry water from the roots to the leaves (transpiration). Phloem cells carry dissolved sugars made during photosynthesis to other parts of the plant (translocation).
- The rate of transpiration is affected by changes in humidity, wind speed, temperature and light intensity.
- Blood comprises red blood cells (which carry oxygen), white blood cells (part of the immune response) and platelets (cell fragments that form scabs). They are carried in blood plasma, a straw-coloured liquid.
- Vaccination results in the production of memory cells, which can produce antibodies sooner, faster and in greater quantity after future exposures to the same antigen.
- Non-specific defence systems against infection include the skin, hairs within the nose, ciliated cells in the trachea and bronchi, and stomach acid.
- If an invading pathogen passes these defences, the immune system attacks. Phagocytes engulf and destroy pathogens. Lymphocytes produce antibodies that 'clump' together pathogens. They also produce antitoxins to neutralise toxins produced by the pathogen.
- Your heart has four chambers. The atria (upper chambers) pump blood to the ventricles. The ventricles then pump it to the body (left ventricle) or lungs (right ventricle).
- The lungs provide the blood with oxygen and remove carbon dioxide.
- Arteries like the aorta and pulmonary artery take blood away from the heart. Arteries then divide into tiny capillaries, which permeate tissues. Veins like the venae cavae and pulmonary vein return blood to the heart.

Exam practice

1 Unlike us, plants do not possess hearts to move substances around their bodies. Instead they must use other processes to ensure all their cells receive the substances they need to survive.

(a) What name is given to the process of moving water from the roots to the leaves of a plant? [1]
    A Transpiration   B Photosynthesis
    C Osmosis   D Translocation

(b) What name is given to the process of moving sugars from the leaves to the rest of a plant? [1]
    A Transpiration   B Photosynthesis
    C Osmosis   D Translocation

(c) Explain how plant roots are adapted for osmosis and active transport, giving examples of a substance that is absorbed by each process. [6]

2 Your heart beats to keep blood moving around your body.

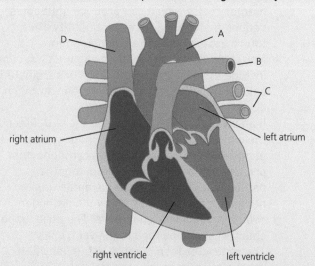

right atrium

left atrium

right ventricle

left ventricle

(a) The diagram shows the human heart. Name its blood vessels A to D. [4]
(b) (i) State which chamber of the heart has the largest muscular lining. [1]
    (ii) Explain why it needs to have such a muscular lining. [1]
(c) Describe how blood passes around the body beginning with the left atrium. Specific names of blood vessels are not required. [6]

3 Pathogens are disease-causing organisms. The common cold is an example.

(a) State which component of a common cold's pathogen antibodies will bind with. [1]
(b) Describe how antibodies and enzymes are similar. [2]

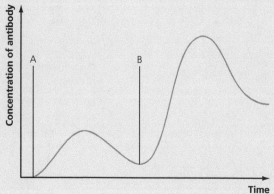

(c) Vaccinations can help prevent serious illness. The graph shows how an immune system responded after vaccination. Write labels for the two points on the graph shown by the letters A and B. [2]
(d) Explain the shape of this graph. [4]
(e) Describe how your non-specific defence systems help to stop you from falling ill with diseases such as the common cold. [6]

Answers and quick quizzes online

ONLINE

4 Coordination and control

Excretion in flowering plants

Excretion is the removal of waste substances from an organism. This is an essential process and without it, an organism will often die relatively quickly.

Plants complete respiration at all times and photosynthesis in the light. The word equation for aerobic respiration is:

$$\text{glucose + oxygen} \xrightarrow{\text{energy out}} \text{carbon dioxide + water}$$

We can see from this that plant cells excrete carbon dioxide and water as a result of respiration.

The word equation for photosynthesis is:

$$\text{carbon dioxide + water} \xrightarrow{\text{light in}} \text{glucose + oxygen}$$

We can see from this that plant cells excrete oxygen as a result of photosynthesis. Plants use glucose as an energy source. The gases are exchanged through tiny pores called stomata, which are found mainly on the underside of leaves.

> **Excretion:** The removal of waste products from an organism.

Now test yourself

TESTED

1 Define the term excretion.
2 What are the waste products from plant respiration?
3 What is the waste products from photosynthesis?

Answers on p. 126

Excretion in humans

The cells in your body are respiring at all times. You produce carbon dioxide and water as a result. These products travel in your blood to your lungs where they are then excreted when you exhale. You also lose water and salt ions through your skin as you sweat.

A final way that you excrete waste is in urine. Your kidneys make urine from the urea and other waste substances in your blood, which can then be expelled. Urea is formed when your body breaks down excess protein. Your lungs, skin and kidneys are therefore all organs of excretion.

The urinary system

REVISED

You have two kidneys, one either side of your spine in your lower back. They are supplied with blood by your renal arteries. After blood is filtered in your kidneys, it returns to the rest of your body via renal veins. Urine produced by your kidneys is sent to the bladder in ureters. It is held in your bladder until you urinate, when the urine is removed from your bladder via your urethra.

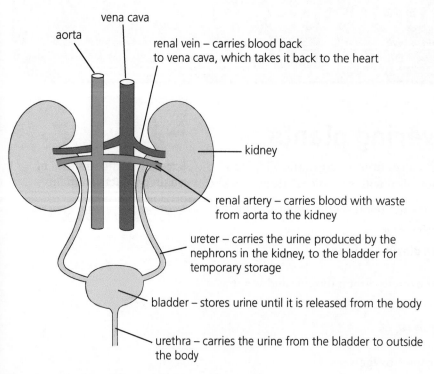

vena cava

aorta

renal vein – carries blood back
to vena cava, which takes it back to the heart

kidney

renal artery – carries blood with waste
from aorta to the kidney

ureter – carries the urine produced by the
nephrons in the kidney, to the bladder for
temporary storage

bladder – stores urine until it is released from the body

urethra – carries the urine from the bladder to outside
the body

Figure 4.1 **The key parts of your urinary system**

The nephron

Within each of your kidneys are about a million smaller structures called **nephrons**. The nephrons remove excess water, ions and urea from your blood before it leaves the kidney in your renal veins.

This process in your kidneys has three main steps:
1 All glucose, urea, ions and water are removed from the blood (ultrafiltration).
2 All glucose, and some ions and water, are selectively reabsorbed into the blood (selective reabsorption).
3 All urea, and excess ions and water, are excreted in your urine (excretion).

Ultrafiltration occurs in the **Bowman's capsule** which surrounds the glomerulus. The glomerulus filters the blood and produces a solution called **glomerular filtrate**. Selective reabsorption of glucose (it is only the glucose molecules that are reabsorbed) occurs in the **proximal convoluted tubule** by active transport. Water is reabsorbed into the blood by osmosis from the collecting duct.

All glucose is therefore kept for respiration and all urea is removed. The appropriate ions and water are retained to maintain optimum levels. This means your urine contains all the urea you produce, as well as some water and ions, but no glucose.

Nephron: The functional unit of the kidney where excess water, ions and urea are removed from the blood.

Bowman's capsule: A cup-shaped structure that surrounds the glomerulus in the nephron and is involved in ultrafiltration.

Glomerular filtrate: The solution produced in the glomerulus of each nephron in your kidneys. It contains water, glucose, salt ions and urea.

Proximal convoluted tubule: The twisted part of the nephron between the Bowman's capsule and the loop of Henle (or nephron loop) where selective reabsorption occurs.

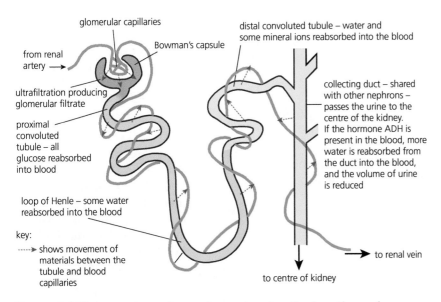

glomerular capillaries

Bowman's capsule

from renal artery →

distal convoluted tubule – water and some mineral ions reabsorbed into the blood

ultrafiltration producing glomerular filtrate

proximal convoluted tubule – all glucose reabsorbed into blood

collecting duct – shared with other nephrons – passes the urine to the centre of the kidney. If the hormone ADH is present in the blood, more water is reabsorbed from the duct into the blood, and the volume of urine is reduced

loop of Henle – some water reabsorbed into the blood

key:
·····► shows movement of materials between the tubule and blood capillaries

→ to renal vein

to centre of kidney

Figure 4.2 **The structure of a nephron showing the locations of ultrafiltration, selective reabsorption and excretion**

Typical mistake

Students often mix up which of the four substances (glucose, urea, ions and water) are absorbed by different processes (ultrafiltration, selective reabsorption and excretion) and where they are absorbed. Make detailed notes on this to ensure you don't get confused in the exam.

Exam tip

Exam questions often ask you to describe the function of the kidneys in maintaining water balance in the body. You should also be able to translate tables and bar charts of glucose, ions and urea before and after filtration.

ADH and osmoregulation

REVISED

Osmoregulation is an example of homeostasis (see page 60). It is the regulation of the correct water concentration within your cells. Your body's water concentration is monitored by the osmoregulatory centre of your brain. The pituitary gland releases **anti-diuretic hormone (ADH)** into your blood, which is transported to the kidneys to control the amount of urine being excreted and stop you becoming dehydrated. Too much water in your body means less ADH is produced. This means you urinate more and your urine is more dilute. Too little water in your body means more ADH is produced. You then pass a smaller volume of more concentrated urine.

Osmoregulation: The regulation of the correct water concentration within your cells.

Anti-diuretic hormone (ADH): A hormone produced in your pituitary gland that regulates the volume of urine you excrete.

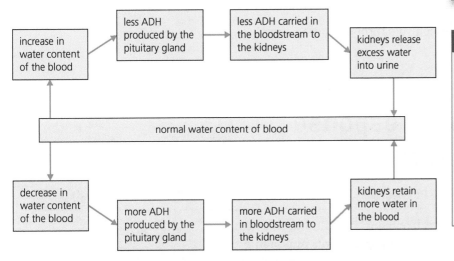

increase in water content of the blood → less ADH produced by the pituitary gland → less ADH carried in the bloodstream to the kidneys → kidneys release excess water into urine

normal water content of blood

decrease in water content of the blood → more ADH produced by the pituitary gland → more ADH carried in bloodstream to the kidneys → kidneys retain more water in the blood

Typical mistake

Students often confuse the ADH hormone with other abbreviations such as FSH and LH, which are involved in the menstrual cycle. Note down any abbreviations that you might easily confuse with their full names and functions so you have a crib sheet for quick revision.

Figure 4.3 **A flow diagram summarising the control of water content in the blood by ADH and the kidneys**

This use of ADH is an example of **negative feedback control**. This type of control occurs when your body detects a change (for example too little or too much water) and tries to correct it in order to return itself to normal.

Now test yourself

TESTED

4 What function do your skin, lungs and kidneys all have in common?
5 What blood vessels supply your kidneys?
6 What is the functional unit of the kidney?
7 What are the names of the three main processes in the kidney?
8 Where in the kidney is water reabsorbed into the blood?
9 What is glomerular filtrate made from?
10 Which substance in glomerular filtrate is completely removed from your body?
11 What does ADH stand for?
12 What does ADH control?

Answers on p. 126

Exam tip

The role of ADH in maintaining water balance in the body often comes up in the exam. You may find it helpful to draw your answer as a flow diagram.

Coordination and response in living organisms

Homeostasis

REVISED

The millions of cells in your body have optimum conditions for functioning. For example, they require enough glucose for respiration, as well as water and an appropriate temperature. **Homeostasis** is the maintenance of these (and many other) optimum conditions. We say that homeostasis is the maintenance of a constant internal environment. These internal changes happen without you knowing so they are automatic (involuntary). They allow your body to meet the challenges of a changing environment.

Negative feedback control is what occurs when your body detects a change, and makes an adjustment to address this and return itself to normal. It is an essential part of homeostasis.

Homeostasis: The maintenance of a constant internal environment.

Exam tip

Make sure you remember the definition of homeostasis shown in the key term box as it's likely to come up in an exam.

Now test yourself

TESTED

13 Define the term homeostasis.
14 Why does your body undergo homeostasis?
15 State three factors your body controls through homeostasis.

Answers on p. 126

Coordination and response in flowering plants

Hormones and plant growth

REVISED

Plants are not able to move positions like many animals, but their shoots can grow towards the light. This is an example of plants responding to a stimulus. Higher light intensities mean plants can perform more photosynthesis, which means more glucose is produced. This in turn means more respiration can occur, which results in more growth. This movement towards light is called **positive phototropism**. The movement of plant roots growing downwards is called **positive geotropism** (or gravitropism). This growth downwards is due to gravity and it helps anchor plants in the ground.

Auxins

Auxins are plant hormones that control phototropism and geotropism. They are produced in the tips of roots and shoots and then they migrate through the cells of the plant to where they are needed. Once in the right place, they cause cells to grow longer in a process called **cell elongation**.

In plant shoots, auxins are concentrated on the shaded side of the plant. This causes the cells in the shade to elongate, which bends the shoot towards the light. In plant roots that are growing horizontally along the ground, auxins are concentrated on the lower side of the root. This inhibits the growth of these cells, which bends the tip of the root downwards.

Plant roots are also able to detect and grow towards water. This process is called **hydrotropism**.

> **Positive phototropism:** The ability of plant shoots to grow towards the light.
>
> **Positive geotropism:** The ability of plant roots to grow downwards (also called gravitropism).
>
> **Auxin:** A type of plant hormone responsible for cell elongation.
>
> **Cell elongation:** The lengthening of specific cells in plants as a result of hormones.
>
> **Hydrotropism:** An ability of plant roots to grow towards water.

Experiments involving auxins

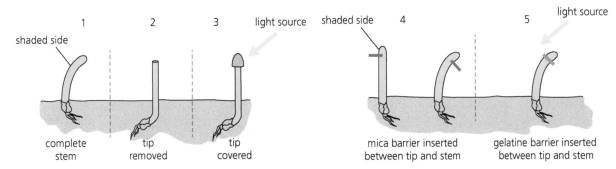

Figure 4.4 A number of experiments have proved that auxins are responsible for phototropism.

Figure 4.4 shows the results of an investigation into phototropism. Table 4.1 explains these results.

> **Impermeable:** Describes something through which a substance cannot pass.

Table 4.1 **An explanation of the results of experiments into phototropism**

Diagram	Experimental details	Results and explanation
1	No changes made	This is a typical response. The shoot is growing towards the light.
2	The tip of the shoot has been removed	No phototropism occurs. This proves that auxins are produced in the tip of the shoot.
3	The tip of the shoot has been covered	No phototropism occurs. This proves that the plant's light sensitive cells are in the tip.
4	Both sides of the tip of the shoot have had a mica (**impermeable**) barrier inserted	No phototropism occurs when the barrier is on the shaded side. Phototropism does occur when the barrier is on the light side. This proves that auxins are concentrated on the shaded side of the shoot.
5	The tip of the shoot has had a gelatine (permeable) barrier inserted	Phototropism occurred. This proves that auxins can diffuse through permeable barriers like gelatine.

Other hormones

Gibberellins are a second group of plant hormones. Like auxins, they have several functions. They help with stem elongation, dormancy before seeds germinate, germination and the formation of flowers and fruit. **Ethene** is another plant hormone, which is responsible for the ripening of fruit.

Uses of plant hormones

Agriculture is the growth of animals or crops for food, fuels or medicines. Horticulture is the growth of plants. We use plant hormones in both agriculture and horticulture.

Auxins can be used as selective weed killers. When sprayed on weeds in grass, more auxin hormones land on the larger leaves of the weeds. This causes some plants to grow uncontrollably (which kills them) and others to stop growing.

Auxins are also used in rooting powder when plant cuttings are taken. This is shown in Figure 4.5. They encourage cells found in the stems of plants to turn into roots.

Auxins are used in tissue culture as well. This is when a small number of cells are removed from a plant to grow it into a genetically identical clone. Auxins are used to promote growth in these removed cells.

Ethene is used by the food industry to ripen fruit before it goes on sale. Much of our fruit is imported and is picked before it is ripe to avoid it rotting on the journey. Gibberellins are used to end seed dormancy, promote flowering in plants and increase the size of fruit.

> ### Now test yourself
> TESTED ☐
>
> 16 What is positive phototropism?
> 17 What is positive gravitropism?
> 18 Name two different types of plant hormone.
> 19 How do plant shoots grow towards the light?
> 20 Define the term cell elongation.
> 21 What is ethene used for in plants?
> 22 Which plant hormone helps with stem elongation, dormancy before seeds germinate, germination and the formation of flowers and fruit?
> 23 State three ways in which humans use plant hormones.
> 24 How do auxins work as selective weedkillers?
> 25 How do plant hormones used in rooting powder help us grow plant cuttings?
>
> Answers on p. 126

Coordination and response in humans

Two major organ systems control coordination and response in your body: the nervous system and the endocrine system.

> **Gibberellins:** Plant hormones responsible for cell elongation, seed dormancy and germination.
>
> **Ethene:** A plant hormone that ripens fruit.

> ### Revision activity
>
> Study Figure 4.4 and write a couple of sentences to explain the results of each experiment. Then check your answers against Table 4.1 to see how accurate your conclusions were.

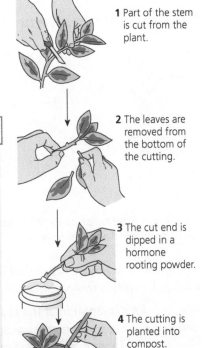

1 Part of the stem is cut from the plant.

2 The leaves are removed from the bottom of the cutting.

3 The cut end is dipped in a hormone rooting powder.

4 The cutting is planted into compost.

Figure 4.5 How to take a plant cutting

> ### Exam tip
>
> Exam questions often ask about how both plants and humans use hormones. Make sure you know examples of both.

The nervous system

Your nervous system controls your voluntary and involuntary actions. It allows you to react to your surroundings and coordinate your behaviour. It is made up of millions of neurones that transmit and receive millions of messages each day. Your nervous system includes:

1 your **central nervous system (CNS)** – your brain and spinal cord
2 your peripheral nervous system – the millions of nerves that criss-cross the rest of your body. Nerves are made from bundles of individual neurones.

Sending electrical impulses

All messages sent along neurones in your nervous system are electrical. These messages travel very quickly. This electrical signal is generated by a cell called a **receptor**. You have these receptor cells in all your sense organs, including your skin. Some areas of your skin like your lips and fingertips are especially sensitive. This is because they have a lot more receptors than other, less sensitive areas.

Table 4.2 **Your senses and the organs and stimuli involved**

Sense	Organ	Stimulus
Sight	Eyes	Light
Hearing	Ears	Sound
Taste	Tongue	Chemicals in food
Smell	Nose	Chemicals in air
Touch	Skin	Touch, pressure, temperature, pain and itch

Sensory, relay and motor neurones

There are three main types of neurone:

1 **Sensory neurones** carry signals from receptors to your CNS.
2 **Relay neurones** carry signals around your CNS.
3 **Motor neurones** carry signals away from your CNS to **effectors**.

> **Typical mistake**
>
> Students often mistake the direction in which electrical signals pass along the three types of neurone. Remember that sensory neurones carry signals to the CNS, relay neurones within it, and motor neurones away from it.

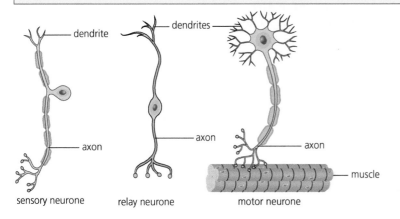

sensory neurone   relay neurone   motor neurone

Figure 4.7 **The three types of neurone**

> **Central nervous system (CNS):** The brain and spinal cord.
>
> **Receptor:** A cell or group of cells at the beginning of a pathway of neurones that detects a stimulus and generates an electrical impulse.

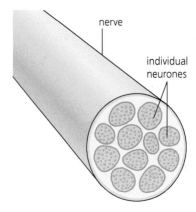

Figure 4.6 **A bundle of neurones is a nerve.**

> **Revision activity**
>
> Draw out Table 4.2 with only the headings. Try to fill in the rest of the table from memory to help you revise.

> **Sensory neurone:** A neurone that carries an electrical impulse from a receptor towards the central nervous system.
>
> **Relay neurone:** A neurone that carries an electrical impulse within the central nervous system.
>
> **Motor neurone:** A neurone that carries an electrical impulse away from the central nervous system to an effector (muscle or gland).
>
> **Effector:** A muscle that moves or a gland that produces a hormone.

Motor neurones end in either muscles or glands. Muscles contract or relax to move parts of your body in response to the signals. Glands respond by releasing hormones into the blood. Both muscles and glands are known as effectors. The pathway of an electrical signal from receptor to effector is:

stimulus ⟶ receptor ⟶ sensory neurones ⟶ motor neurones ⟶ effector ⟶ response

A **coordinated response** therefore requires a stimulus that you detect, a receptor to begin the electrical signal, neurones to carry it to and from your central nervous system, and an effector.

Synapses

There is more than one nerve that links each of your fingers to your brain. These nerves form a network. If one nerve is damaged, the signal can be re-routed and still reach your brain. Each electrical signal passes along multiple nerves. There are gaps between the neurones called **synapses**.

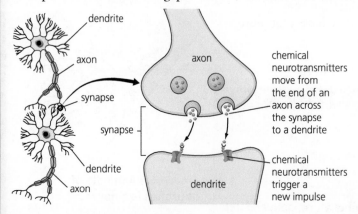

Figure 4.8 How signals move from one neurone to another across a synapse

The ends of neurones branch into **dendrites**. At the tips of these branches are special areas that make chemicals called **neurotransmitters**. When an electrical signal reaches the end of a neurone, it triggers the release of neurotransmitters into the synapse. These chemicals diffuse across the synapse, bind to receptors and restart the electrical signal in the next neurone. This whole process happens extremely quickly.

The reflex arc

Some reactions are automatic. For example, you don't have to think about moving your hand if you touch something hot; it happens instinctively. These reactions are called **reflex responses**. In reflex responses the electrical signal is not initially transferred to the conscious brain. The signal is started by receptors and travels along sensory neurones (as always) to your spinal cord. The relay neurones then immediately send a signal along motor neurones to your muscles, which contract and move your body. The signal is sent to your conscious brain shortly afterwards. This makes the reaction quicker than if it had gone via your brain. This reduces possible damage to your body.

Revision activity

Draw two different flow diagrams to show the differences between a normal and a reflex response in your nervous system. Refer back to these diagrams when revising.

Coordinated response: A response that occurs when your nervous system detects a change. In a coordinated response, an electrical signal is transferred to and from your CNS, which results in a muscle moving or a gland producing a hormone.

Synapse: A gap between the axon of one neurone and the dendrites of another where neurotransmitters transmit the impulse.

Dendrites: The branched beginnings of neurones, which can detect neurotransmitters and start another electrical impulse.

Neurotransmitter: A chemical substance released at the end of one neurone that diffuses across a synapse to begin a second electrical impulse in another neurone.

Reflex arc: The movement of an electrical impulse that avoids the brain to save time and probable damage to your body.

Reflex response: An automatic response that you do not have to think about.

Exam tip

You should be able to explain how the structure of synapses and neurones (sensory, relay and motor) all relate to their function. You should also be able to explain the importance of reflex actions.

The eye

The eye is the organ of vision. It allows you to see by detecting light and converting it into electrical signals that are passed along nerves to your brain.

Structure of the eye

The structure of the eye is shown in Figure 4.9 and the functions of its components are given in Table 4.3.

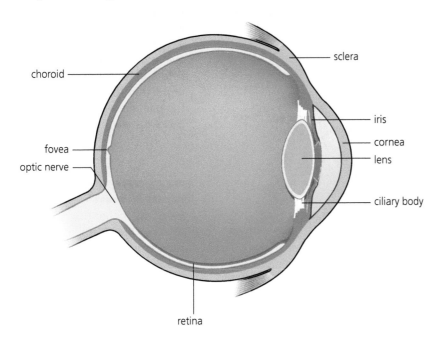

Figure 4.9 **The structure of the mammalian eye in cross-section**

Rods: Light-sensitive receptor cells on your retina in the back of your eye that let you see in low light conditions.

Cones: Light-sensitive receptor cells on your retina in the back of your eye that let you see in colour.

Table 4.3 **The parts of the mammalian eye and their functions**

Part	Function
Cornea	The transparent part of your eye that covers the iris and pupil. It refracts light through the pupil.
Lens	The biconvex structure that sits behind your pupil. It further refracts light to be focused on your retina. Its shape changes in a process called accommodation when you look at near and far away objects.
Iris	The coloured muscle that surrounds your pupil. It controls the size of your pupil by relaxing and contracting to let more or less light in.
Ciliary body	A structure made of muscles called ciliary muscles which can contract and relax to change the shape of your lens during accommodation.
Choroid	The layer of your eye that is found between the retina and sclera. It provides oxygen and nourishment to the cells of the retina.
Retina	The layer of receptor cells found inside your eye. It contains two types of light-sensitive cells called **rods** and **cones**.
Fovea	A specific part of your retina that is the centre of your sharpest vision.
Sclera	The white of your eye. It is the outer layer that protects the rest of your eye. In many other mammals it is not white and so not as easy to distinguish from the iris.
Optic nerve	The nerve made from the sensory and motor neurones that connect your brain to your eyes. Electrical impulses travel along your optic nerve to allow you to see.

4 Coordination and control

Accommodation: The shape of the lens in your eye changing to focus on near or far objects.

Accommodation

The lenses in your eyes can change shape to allow you to see near or far objects. This process is called the **accommodation** reflex and it is automatic. The changes in lens shape during accommodation are shown in Figure 4.10.

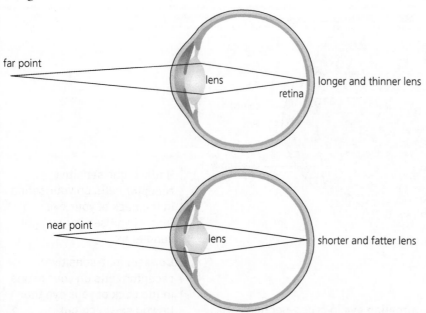

far point

lens

retina

longer and thinner lens

near point

lens

shorter and fatter lens

Figure 4.10 **When you look at closer objects, your lens becomes short and fat in shape. When you look at objects further away, your lens becomes longer and thinner.**

To see a near object, your ciliary muscles contract and the lens becomes thicker. This causes light rays to be refracted more strongly. To see a distant object, your ciliary muscles relax and the lens becomes thinner. This causes light rays to be refracted less.

Thermoregulation

REVISED

Your body functions most effectively between the temperatures of 36.5 and 37.5°C. This is the optimum temperature at which most of your enzymes function. The thermoregulatory centre in your brain controls your body temperature. This centre contains receptors that detect the temperature of the blood. Your skin also contains temperature receptors to detect changes in external temperature. They send electrical signals along neurones to your thermoregulatory centre so that it can respond accordingly.

Sweating and shivering

If your thermoregulatory centre detects that you are too hot, sweat glands in your skin produce sweat. This sweat evaporates from your skin, which transfers energy away to cool you.

When your body becomes too cold you automatically begin to shiver. This is a reflex response that you cannot control. It is caused by your skeletal muscles contracting and relaxing very quickly. This movement generates heat to warm you.

Vasoconstriction and vasodilation

When you are too hot, blood vessels open near the surface of your skin. This means there is more blood nearer your skin's surface and the warm blood is able to lose excess heat. This process is called **vasodilation**. Your cheeks may appear flushed as a result of increased blood at the surface of your skin.

The reverse happens if your thermoregulatory centre detects you are too cold. The same blood vessels near your skin close to reduce the volume of blood that can lose heat. This is called **vasoconstriction**. Your skin often appears paler when you are cold as a result of reduced blood at the surface of your skin.

hairs standing up to keep moving air away from skin and trap a layer of warm air

muscle pulls hair erect

less active sweat gland

hairs flatten and allow moving air closer to skin

droplets of sweat evaporate

active sweat gland

vasoconstriction of arterioles near skin surface diverts blood to vessels further away from skin to retain heat

vasodilation of arterioles near the skin surface, allows more blood to flow in the capillaries close to the skin surface to lose heat

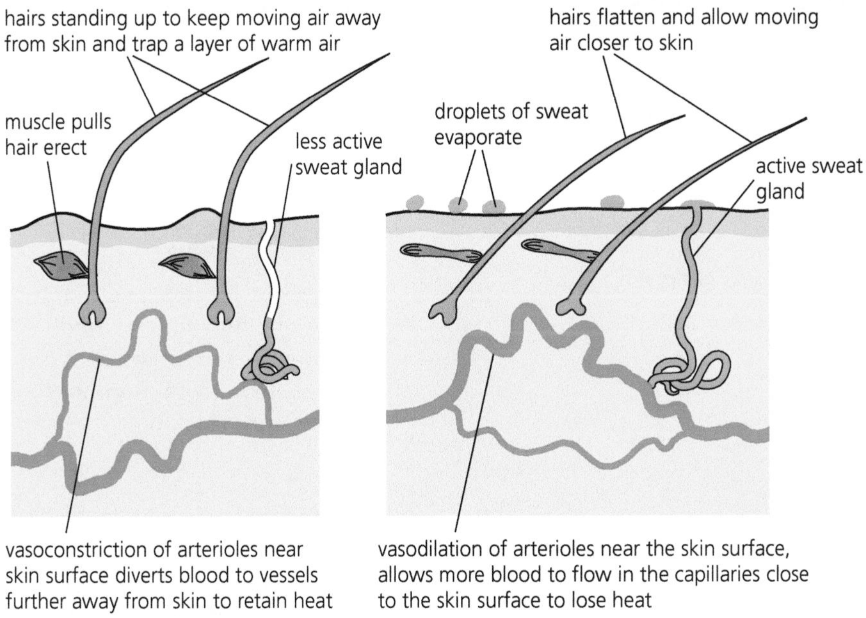

Figure 4.11 **Vasodilation and vasoconstriction**

Hypothermia and hyperthermia

A body temperature of between 37.5 and 38.5 °C is called a fever. Above 38.5 °C **hyperthermia** begins. This increase in temperature is often caused by heat stroke or an allergic reaction to drugs. Without a reduction in temperature, hyperthermia can be fatal.

A body temperature below 35 °C is called **hypothermia**. This often results in mental confusion and people with hypothermia may actually try to take their clothes off. Without an increase in temperature, hypothermia is fatal.

The endocrine system

REVISED

Your **endocrine system** is a group of glands that secretes hormones into your blood. These glands, their hormones and their functions are shown in Table 4.4 and Figure 4.13. Hormones travel in the blood and so hormonal responses are usually slower than those of the nervous system.

Vasodilation: The increase in size of blood vessels so more blood flows at the surface of the skin to increase heat loss.

Vasoconstriction: The decrease in size of blood vessels so less blood flows at the surface of the skin to reduce heat loss.

Hyperthermia: A medical condition caused by high body temperature, which causes organ failure and eventually death.

Hypothermia: A medical condition caused by low body temperature, which causes organ failure and eventually death.

Typical mistake

Students often confuse the terms hypothermia and hyperthermia. Remembering that 'hypo...' rhymes with 'low' (temperature) is a useful way of remembering.

Endocrine system: The system of glands that secrete hormones into the circulatory system.

Table 4.4 Common examples of hormones and their functions

Hormone	Location	Target organ(s)	Function
Adrenaline	Adrenal glands	Heart and other vital organs	Prepares the body for 'fight or flight'
Insulin	Pancreas	Liver and muscles	Reduces blood sugar levels
Testosterone	Testes	Reproductive organs	Controls puberty in men
Progesterone	Ovaries	Reproductive organs	Controls puberty and the menstrual cycle in women
Oestrogen	Ovaries	Reproductive organs	Controls puberty and the menstrual cycle in women
ADH (anti-diuretic hormone)	Brain (made in hypothalamus and stored in pituitary gland)	Kidneys	Controls amount of water in urine (this is covered in detail on page 59)
FSH (follicle-stimulating hormone)	Brain (pituitary gland)	Ovaries	Causes an ovum to mature inside a follicle in the ovary and stimulates the ovary to produce oestrogen
LH (luteinising hormone)	Ovary	Ovaries	Stimulates ovulation (the release of an ovum from an ovary)

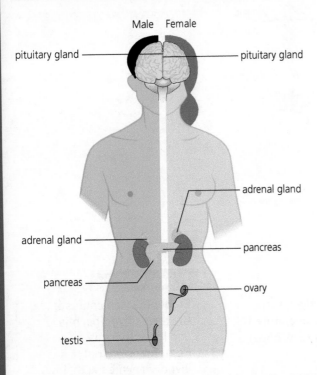

Figure 4.13 **The positions of key glands in your endocrine system**

You will learn more about testosterone, progesterone and oestrogen in Chapter 5 (from page 71).

Insulin

When your blood glucose level is too high, your pancreas releases insulin. This hormone travels in the bloodstream to your liver and muscles and causes your cells to convert excess glucose into insoluble glycogen. This glycogen is then stored in your liver and muscle cells so that your blood glucose levels are reduced and return to normal.

> **Revision activity**
>
> Draw out Table 4.4 with only the headings and hormone names. Try to fill the rest of the table in from memory to help your revision.

> **Exam tip**
>
> Exam questions often ask you to identify the positions of the glands in Figure 4.13, so it is important that you know their location and function.

> Insulin: A hormone produced in your pancreas that lowers blood glucose by converting it to glycogen and storing it in the liver.

> **Exam tip**
>
> You should be able to explain how insulin controls blood glucose levels.

Pituitary gland

The **pituitary gland** in your brain is your 'master gland'. It secretes hormones that directly control growth and blood pressure. It also partly controls how your ovaries or testes and your kidneys function, as well as pregnancy and childbirth. It does this by releasing the hormones that control other glands and organs.

> **Pituitary gland:** A gland in your brain that produces growth hormones, as well as ADH, FSH (only in women) and LH (again only in women).

Now test yourself

26 What makes up the central nervous system?
27 What are effectors?
28 Describe the pathway of an electrical signal from receptor to effector.
29 How is the pathway of a reflex action different from a 'normal' nervous response?
30 How does the lens change shape during accommodation when viewing a distant object?
31 What occurs during vasodilation?
32 What medical condition occurs when a person's body becomes too hot?
33 In which type of structures are hormones made?
34 What happens when the blood glucose level is too high?
35 What hormone controls the volume of urine? Where is this hormone produced?

Answers on p. 126

> **Revision activity**
>
> Draw a mind map of all the content in this chapter. You could use the topic headings: excretion in flowering plants, excretion in humans, coordination and response in living organisms, coordination and response in flowering plants, and coordination and response in humans.

Summary

- Excretion is the removal of waste substances.
- Plants excrete carbon dioxide and water as a result of respiration. They also excrete oxygen as a result of photosynthesis. These gases are excreted through stomata.
- Your cells excrete carbon dioxide and water as a result of respiration. This occurs in your lungs. Your skin excretes water and salt ions when you sweat. Your kidneys make urine from urea and other waste substances. This means your lungs, skin and kidneys are all organs of excretion.
- All glucose, urea, ions and water are removed from the blood (ultrafiltration). All glucose, some ions and some water are selectively reabsorbed into the blood (selective reabsorption). All urea, excess ions and excess water are excreted in your urine (excretion).
- ADH is produced in the pituitary gland in the brain. It controls the volume of water in your urine. Too little water in your body means you produce more ADH. This means you pass less water in your urine. The reverse also applies.
- Homeostasis is the maintenance of a constant internal environment. These conditions are optimum for enzymes and all cell functions. Homeostasis includes controlling blood glucose concentration, body temperature and water levels.

- Auxins are plant hormones that are formed in the tips of shoots to help plants grow towards the light (phototropism) and in the tips of roots to help plants grow downwards (gravitropism or geotropism). In both processes, auxins cause some cells to become longer (elongation).
- The nervous system enables humans to respond to their surroundings and coordinate behaviour. Information from receptor cells passes along sensory neurones as electrical impulses to the central nervous system (CNS). The CNS is the brain and spinal cord. Within the CNS relay neurones transmit electrical impulses. Motor neurones carry electrical signals to effectors.
- Synapses are gaps between neurones that allow one neurone to pass a signal to others.
- In reflex reactions, signals are not immediately transferred to the conscious part of the brain. This saves time and potential damage to the body.
- The eye is a sense organ that contains receptors sensitive to light. The eye contains the following parts: retina, optic nerve, sclera, cornea, iris, ciliary muscles and suspensory ligaments.
- The endocrine system is composed of glands that release hormones into the blood. Hormones are carried to target organs where they then act. These hormones are usually slower than nervous system responses.

→

● Adrenaline boosts heart rate and the delivery of oxygen to the brain and muscles to prepare for a 'fight or flight' response. Insulin reduces the amount of glucose in your blood. Testosterone controls puberty in men. Oestrogen and progesterone control the menstrual cycle and puberty in women.

Exam practice

1 Hormones are often thought of as chemical messages. Both animals and plants possess hormones.

(a) Which of these is **not** a use of plant hormones? [1]

   A   Selective insecticides
   B   Producing seedless fruit
   C   Rooting powder
   D   Fruit ripening

(b) Describe what the results show for the experiments shown in the three plant diagrams on the right [3]

2 Your endocrine system possesses many key glands. These produce hormones that are released into your blood.

(a) Give the names of the glands A to D on the image on the right. [4]

(b) The table below shows some of the important hormones found in humans. Fill in the missing sections by copying and completing the table. [3]

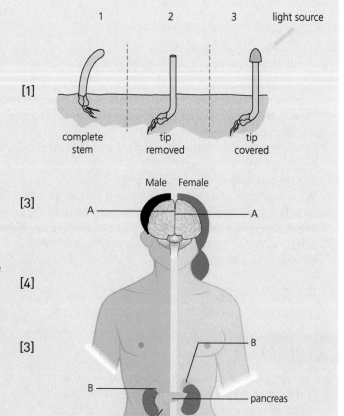

Hormone	Location	Target organ(s)	Function
ADH (anti-diuretic hormone)		Kidneys	Controls amount of water in urine
FSH (follicle-stimulating hormone)	Brain (pituitary gland)	Ovaries	
	Ovary	Ovaries	Stimulates ovulation (the release of an ovum from an ovary)

(c) Hormones play an important role in homeostasis. Define homeostasis. [1]

(d) Describe how your body controls its temperature. Give examples of what happens when you are too hot and too cold in your answer. [6]

3 The eye is a sense organ responsible for sight.

(a) Explain the process of accommodation in the eye. [4]

(b) There are two types of cell found in the retina – rods and cones. State their functions. [2]

(c) Your vision is partially a result of signals travelling along nerves in your nervous system. Describe the structure and function of your nerves and nervous system. [6]

Answers and quick quizzes online

ONLINE

5 Reproduction and inheritance

Reproduction in living organisms

Sexual and asexual reproduction

Asexual reproduction

Asexual reproduction involves one parent organism that produces offspring that are genetically identical to itself. These offspring are called clones and they are clones of each other as well as their one parent. This is because there is no joining of gametes (sex cells), so there is no mix of DNA. Instead the process involves mitosis (see pages 86–7).

Many plants reproduce asexually in a process called vegetative reproduction. All bacteria reproduce asexually when they divide by binary fission, a process similar to mitosis.

Advantages of asexual reproduction include the following:
- Only one parent is needed.
- It is a more time- and energy-efficient process as finding a mate is not required.
- It is faster than sexual reproduction.
- Many identical offspring can be produced at a time when conditions are favourable.

Sexual reproduction

Sexual reproduction involves two parents that produce genetically different offspring. The parent organisms produce gametes, which then fuse during fertilisation. This forms a zygote that undergoes mitosis and develops into an embryo.

Gametes are ova and sperm in animals, or ova and pollen in flowering plants. In both types of organism there is a mixing of genetic information that leads to variety in offspring.

Advantages of sexual reproduction include the following:
- Production of genetic variation in offspring. If the environment changes, this variation may give a survival advantage by natural selection.
- Natural selection can be sped up by humans in selective breeding to increase food production. You will learn more about this in Chapter 7.

The formation of gametes during meiosis (see pages 87–8) requires energy. Some animals and plants produce gametes in incredibly high numbers. For example, many fish release millions of gametes at one time. The process of finding a mate (courtship) also often requires energy. Organisms that reproduce asexually do not need to use this energy.

Both sexual and asexual reproduction

Some organisms can reproduce both sexually and asexually depending upon the circumstances. Examples include:
- the malarial parasite, which reproduces asexually in a human host but sexually inside a mosquito

---

Asexual reproduction: Reproduction involving one parent that creates genetically identical offspring.

Gametes: Sex cells, for example sperm, ova and pollen.

Mitosis: Cell division that involves the formation of identical diploid cells.

Binary fission: The asexual reproduction of bacteria.

Genetic variation: Inherited differences between organisms.

Natural selection: The process by which organisms that are better adapted are more likely to survive and reproduce.

Meiosis: Cell division that involves the formation of four non-identical haploid gametes (sex cells).

Courtship: Behaviours to attract a mate.

Parasite: An organism that damages its host but also depends on it to survive.

Typical mistake

Students often confuse which advantages and disadvantages belong to asexual reproduction and which belong to sexual reproduction. It is important you can remember which is which.

- many fungi, which can reproduce asexually by producing mitotic spores and sexually by producing meiotic spores
- many plants, which produce seeds sexually but can also reproduce asexually during vegetative reproduction.

Reproduction in flowering plants

Flower structure

REVISED

The flower is the reproductive organ of flowering plants. A generalised structure of an insect-pollinated flower is shown in Figure 5.1. The functions of the flower parts are shown in Table 5.1.

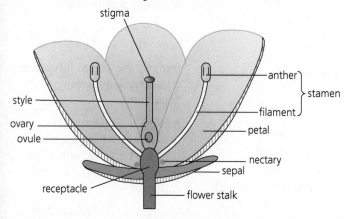

Figure 5.1 **The generalised structure of an insect-pollinated flower**

Table 5.1 **The functions of the parts of an insect-pollinated flower**

Part of flower	Function
Stem or flower stalk	Supports the flower in an appropriate position that will attract insect pollinators.
Sepals	Enclose the flower when it is developing (in bud) for protection.
Petals	Often brightly coloured to attract insects for pollination (if insect pollinated). May be smaller and less colourful if pollinated by wind.
Stamen	This is made from the anther and filament and is the male part of the flower.
Anther	Contains the pollen, which is the male gamete.
Filament	Supports the anther in a position that will attract a pollinating insect.
Carpel	This is made from the stigma, style and ovary and is the female part of the flower.
Stigma	Receives pollen from pollinating insects.
Style	Supports the stigma to receive pollen. The pollen tube develops down the style to reach the ovary.
Ovary	Contains the ovules.
Ovule	The female gamete.

Wind-pollinated flowers are often smaller and less colourful than insect-pollinated flowers. A generalised structure of a wind-pollinated flower is shown in Figure 5.2. The functions of the parts are the same as in Table 5.1.

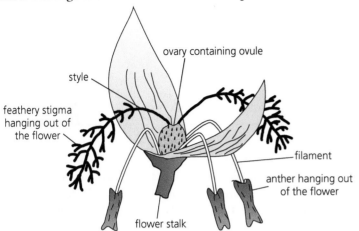

Revision activity

Draw out Table 5.1 with only the headings and the first column. Try to fill the functions in from memory to help you remember them.

Figure 5.2 **The generalised structure of a wind-pollinated flower**

Pollination

REVISED

The female gamete (sex cell) in plants is the **ovule**. This is found in the ovary, which is present in the lower half of the flower (of flowering plants). The male gamete is **pollen**, which is found in the anther. **Pollination** occurs when pollen is transferred to the stigma of a flower. If this is the same flower that produced the pollen, or another flower on the same plant, this is called self-pollination because only one plant is involved. The transfer of pollen to a flower of a different plant is called cross-pollination. Because gametes are involved, both self-pollination and cross-pollination are forms of sexual reproduction.

Ovule: The female sex cell (gamete) of plants.

Pollen: The male sex cell (gamete) of plants.

Pollination: The transfer of the male gamete (pollen) during sexual reproduction in flowering plants.

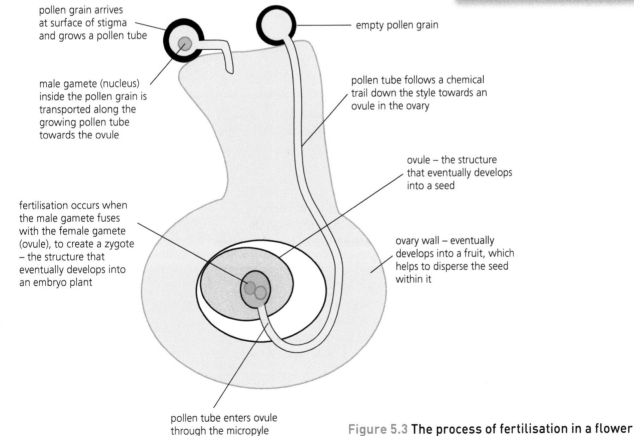

Figure 5.3 **The process of fertilisation in a flower**

When a pollen grain is transferred to the stigma of the same or a different flower, this starts the growth of a pollen tube. These structures grow down the plant's style until they reach the ovule inside the ovary. The nucleus of the pollen grain (which contains the DNA) then moves down the pollen tube and fertilises (joins with) the nucleus of the ovule.

Shortly after fertilisation, the female parts of the plant develop further. The ovules grow into seeds and the rest of the carpel becomes the fruit.

Wind pollination and insect pollination

REVISED

Flowering plants are usually pollinated either by insects or the wind. Insect-pollinated flowers attract insects by having large, colourful flowers and nectar for them to consume. Wind-pollinated plants like grasses have smaller, less colourful petals because they do not need to attract insects. Their stamens and stigmas hang outside the flower. This makes it easy for pollen to be blown from the male anther and collected by the female stigma.

> **Nectar**: A sugar-rich substance produced by flowers to attract pollinating insects.

A summary of some of the adaptations of wind- and insect-pollinated flowers is shown in Table 5.2.

Table 5.2 **Adaptations of insect- and wind-pollinated flowers**

Feature	Wind-pollinated flowers	Insect-pollinated flowers
Petals	Small, less colourful	Large, brightly coloured
Scent	No	Yes – to attract insects
Nectar	No	Yes – to attract insects
Pollen grains	Larger numbers of smaller, lighter grains to be blown in wind	Smaller numbers of spiky or sticky grains to attach to insects
Anthers	Outside flower	Inside flower
Stigma	Outside flower	Inside flower

> **Revision activity**
>
> Make a set of flash cards with a feature of a flower on one side of each card, and its role in plant reproduction on the back.

Seed dispersal and germination

REVISED

Seeds are often dispersed away from the parent plant to avoid competition. The dispersal can be by wind, water or animals (which the seeds either stick to or are eaten by). Other seeds are simply ejected from their pods as they dry.

> **Dispersal**: The spreading of seeds by wind, water, animals or ejection.
>
> **Germination**: The growth of a plant from a seed.

Seeds are often eaten by animals such as birds because they are surrounded by fruits. These animals consume the fruit as an energy source and often eat the seeds too. These seeds then pass through the animal's digestive system unharmed and often germinate in the place they are defecated, as they have ready-made manure as a fertiliser.

Germination is the development of a plant from a seed. Seeds will only germinate and start to grow under the correct conditions. These conditions include the presence of water, oxygen and warmth. The embryo of a seed contains the immature root and shoot, and a seed also possesses a seed coat for protection and its own food store, which will provide energy until the plant can begin photosynthesis.

Required practical 1

Investigate the conditions needed for seed germination

Equipment:
- Cotton wool
- Four boiling tubes
- Cress seeds
- Water

Method:
- Cotton wool was placed into the bottoms of four boiling tubes.
- Ten cress seeds were placed on top of each piece of cotton wool.
- The first tube was left dry and placed at room temperature in the light.
- The second tube had water added and was kept at room temperature in the light.
- The third tube had water added and was kept at room temperature in the dark.
- The last tube had water added and was kept at a colder temperature in the dark.
- All tubes were left for five days, before the number of germinated seeds was counted.
- The results were compared.

Results:
- The results are shown in Figure 5.4.
- The seeds germinated best with water, at room temperature and in the light (tube B).
- The seeds without water (tube A) and in the cold and dark (tube D) did not germinate at all.
- The seeds did germinate with water and warmth in the dark (tube C) but without light they did not grow healthily, and looked yellow.

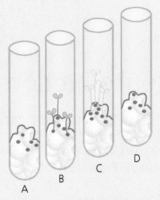

Figure 5.4 **The results of an investigation into the conditions needed for seed germination**

Asexual reproduction in plants

You have already learned that asexual reproduction involves one parent organism that produces genetically identical offspring called clones. There is no joining of gametes (sex cells), so there is no mix of DNA. Many plants like strawberries and spider plants reproduce asexually. These plants form offshoots called **runners**, which 'run' along the ground away from the plant. At regular intervals on these runners, small plants begin to form. These are called plantlets and are clones of the parent plant (so are genetically identical). The process of forming plantlets on runners is an example of vegetative reproduction.

The advantages of this form of asexual reproduction for plants include the fact that only one parent is needed and it is quicker and uses less energy than sexual reproduction. More details about the advantages of asexual reproduction are found earlier in this chapter (see pages 71-2).

Asexual reproduction is also used by gardeners to reproduce plants with desirable characteristics. Gardeners achieve this by taking cuttings or through **micropropagation** of plants. These are both types of artificial asexual reproduction, as they would not happen naturally. To improve the success of this process, gardeners will often use rooting powder that contains plant hormones.

Runner: An offshoot of a plant on which plantlets are produced through asexual reproduction.

Micropropagation: The cloning of plants with desirable characteristics by taking a small sample of tissue from the parent plant and growing it in a sterile medium.

TESTED

Now test yourself

6 What is the function of sepals in a flower?
7 What are the male parts of the flower?
8 What are the female parts of the flower?
9 Describe the process of pollination.
10 How do insect-pollinated flowers attract pollinators?
11 Give two ways that wind-pollinated flowers are different from insect-pollinated ones.
12 Define the term dispersal (relating to seeds).
13 What conditions are needed for germination?
14 How do strawberries and spider plants reproduce asexually?

Answers on p. 126–7

Reproduction in humans

The female reproductive system

REVISED

The parts of the female reproductive system and their functions are shown in Figure 5.5.

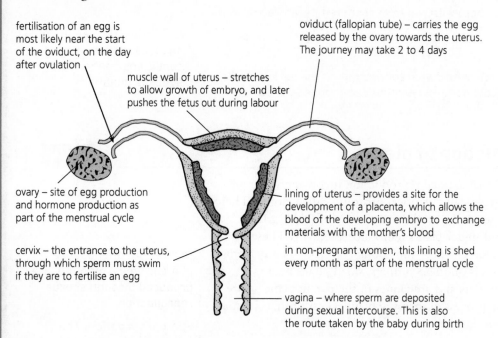

fertilisation of an egg is most likely near the start of the oviduct, on the day after ovulation

muscle wall of uterus – stretches to allow growth of embryo, and later pushes the fetus out during labour

oviduct (fallopian tube) – carries the egg released by the ovary towards the uterus. The journey may take 2 to 4 days

ovary – site of egg production and hormone production as part of the menstrual cycle

lining of uterus – provides a site for the development of a placenta, which allows the blood of the developing embryo to exchange materials with the mother's blood

in non-pregnant women, this lining is shed every month as part of the menstrual cycle

cervix – the entrance to the uterus, through which sperm must swim if they are to fertilise an egg

vagina – where sperm are deposited during sexual intercourse. This is also the route taken by the baby during birth

Figure 5.5 **The parts of the female reproductive system and their functions**

The male reproductive system

REVISED

The parts of the male reproductive system and their functions are shown in Figure 5.6.

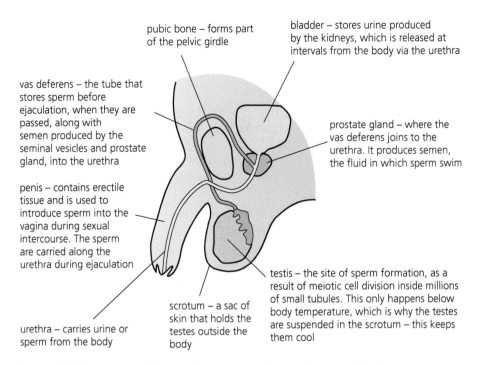

pubic bone – forms part of the pelvic girdle

bladder – stores urine produced by the kidneys, which is released at intervals from the body via the urethra

vas deferens – the tube that stores sperm before ejaculation, when they are passed, along with semen produced by the seminal vesicles and prostate gland, into the urethra

prostate gland – where the vas deferens joins to the urethra. It produces semen, the fluid in which sperm swim

penis – contains erectile tissue and is used to introduce sperm into the vagina during sexual intercourse. The sperm are carried along the urethra during ejaculation

testis – the site of sperm formation, as a result of meiotic cell division inside millions of small tubules. This only happens below body temperature, which is why the testes are suspended in the scrotum – this keeps them cool

scrotum – a sac of skin that holds the testes outside the body

urethra – carries urine or sperm from the body

Figure 5.6 **The parts of the male reproductive system and their functions**

The menstrual cycle

From puberty to the **menopause**, women undergo a 28-day reproductive cycle called the menstrual cycle. Unless a woman is pregnant, the menstrual cycle begins on day one with the breakdown of the lining of the uterus from the previous cycle. This is called **menstruation**, or having a period, and lasts for a few days. This can be painful and may cause cramps. Shortly after this, the hormone **oestrogen** is produced by the ovaries, which causes the lining of the uterus to start to thicken again. This is a result of the uterus preparing itself for a fertilised ovum (egg) to settle and for the woman to become pregnant.

Around day 14 of the cycle, a fertilised ovum is released from an ovary. This is called ovulation. In the days following ovulation a woman is at her most fertile. If sperm are ejaculated into her vagina, they can swim upwards past the cervix and through the uterus to meet the ovum in a fallopian tube.

If the sperm fertilises the ovum, it is likely to settle into the lining of the uterus and develop into a baby. If this happens, levels of the hormone **progesterone** stay high in the woman's body. This stops her from having another period, which would result in a natural abortion. If the ovum is not fertilised, then progesterone levels drop towards the end of the cycle and the woman menstruates again.

> **Menopause:** The point in a woman's life, usually between 45 and 55 years old, when she stops menstruating and therefore cannot become pregnant.
>
> **Menstruation:** Having a period as a part of the menstrual cycle.
>
> **Oestrogen:** A female sex hormone produced in the ovaries that controls puberty and prepares the uterus for pregnancy.
>
> **Progesterone:** A female sex hormone produced in the corpus luteum in an ovary that prepares the uterus for pregnancy.

Hormones in the menstrual cycle

There are four hormones that work together in the menstrual cycle. They are described in Table 5.3 and their effects are shown in Figure 5.7.

Table 5.3 **The main hormones involved in the control of the menstrual cycle**

Hormone	Released by	Target organ	Effect
Follicle-stimulating hormone (FSH)	Pituitary gland	Ovary	● Causes an ovum to mature inside the follicle in the ovary ● Stimulates ovaries to produce oestrogen
Oestrogen	Ovaries	Uterus	● Causes lining to thicken in the first half of the cycle ● High oestrogen concentration switches off the release of FSH and switches on the release of LH
Luteinising hormone (LH)	Pituitary gland	Ovary	● Stimulates ovulation (the release of the ovum from the ovary)
Progesterone	Ovaries (corpus luteum)	Uterus	● Maintains thick uterus lining if fertilised ovum implants ● Higher concentrations of progesterone generated in pregnancy stop the cycle

Revision activity

Write out the names of the four hormones in Table 5.3 in the four corners of a piece of paper. Write out the effects of the hormones on different sticky notes. Mix up the sticky notes and then test yourself by matching the hormones to their effects.

Follicle: The structure in an ovary where an ovum matures.

Corpus luteum: The empty follicle turns into a corpus luteum after ovulation and releases progesterone.

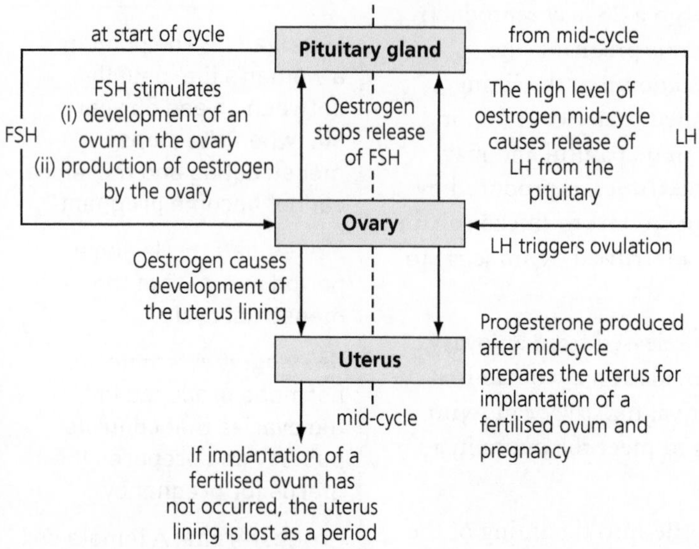

Figure 5.7 **Hormones control the menstrual cycle.**

Revision activity

Draw a flow diagram to show the phases of the menstrual cycle. Include the role of hormones.

Exam tip

Exam questions often ask about the roles of hormones in human reproduction, including the menstrual cycle. Make sure you are clear about the sequence of hormones in the menstrual cycle and their effects. If you are short of time in the exam, writing a flow diagram could help you get down the main points quickly.

Gestation

Gestation is the growth of a baby in the uterus (womb). This is therefore the period of time between fertilisation and birth. It is mainly found in mammals like ourselves, but also occurs in sharks and some invertebrates such as scorpions and worms. The length of gestation in animals differs according to their size; smaller animals such as mice tend to have shorter gestation periods (19 days), whilst larger mammals such as elephants have longer gestation periods (645 days or nearly two years). Gestation in humans is nine months (270 days).

Gestation: The growth of a baby in the uterus.

In mammals, a new organ called the placenta develops in the uterus of the female. This joins the mother to the baby via the umbilical cord. The blood of the baby and mother do not actually mix, but substances are transferred between them by diffusion through the placenta. This is important as the baby requires oxygen and nutrients for growth and it also needs carbon dioxide and other waste products to be removed. Some harmful chemicals, including drugs, can also pass from mother to baby in this way.

During gestation, the baby is surrounded by amniotic fluid in an amniotic sac, which protects the baby. One of the early stages of giving birth is for a woman's 'waters to break' – this actually means that this sac has ruptured and the amniotic fluid is passing from her vagina.

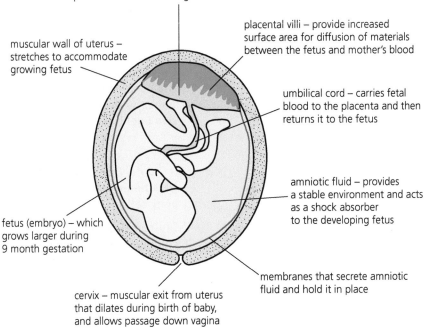

placenta – site of exchange of materials between fetus and mother's blood

placental villi – provide increased surface area for diffusion of materials between the fetus and mother's blood

muscular wall of uterus – stretches to accommodate growing fetus

umbilical cord – carries fetal blood to the placenta and then returns it to the fetus

amniotic fluid – provides a stable environment and acts as a shock absorber to the developing fetus

fetus (embryo) – which grows larger during 9 month gestation

membranes that secrete amniotic fluid and hold it in place

cervix – muscular exit from uterus that dilates during birth of baby, and allows passage down vagina

Figure 5.8 **The parts and functions of the female body during gestation**

Puberty

The sex hormone in men is **testosterone**, which is secreted from the testes. In women, it is the hormones oestrogen and progesterone, which are produced by the ovaries. These hormones are responsible for puberty in both sexes. Oestrogen and progesterone also regulate the menstrual cycle.

Testosterone: A male sex hormone produced in the testes that controls puberty.

Puberty begins for girls at around 11 and for boys usually a year later. Puberty takes about four years to finish and during this time secondary sexual characteristics will also develop. These characteristics are shown in Table 5.4. Puberty does not follow a timetable; it occurs at different times in different people.

Table 5.4 **Secondary sexual characteristics that develop during puberty**

Female	Male
Breasts develop	Testes increase in size
Pubic hair starts to grow	Pubic hair starts to grow
Begins having periods (often around two years after puberty begins)	Voice 'breaks' and becomes deeper
Underarm hair begins to grow	Underarm hair begins to grow
Growth spurt	Growth spurt
Body shape changes (hips widen)	Body shape changes (shoulders widen)
	Facial hair begins to grow

Now test yourself

TESTED

15 Between what stages in their lives do women usually have periods?
16 What does oestrogen control?
17 State the day number of the menstrual cycle around which ovulation usually occurs.
18 What do high levels of progesterone at the end of the menstrual cycle mean?
19 What four hormones control the menstrual cycle?
20 How long is gestation in humans?
21 What is the purpose of amniotic fluid?
22 What hormone controls puberty in men?
23 State a secondary sexual characteristic that only occurs in men.
24 State a secondary sexual characteristic that only occurs in women.

Answers on p. 127

> **Revision activity**
>
> Draw out Table 5.4 with only the headings. Try to fill in the rest of the table from memory to help you to revise.

> **Typical mistake**
>
> Mood swings are sometimes listed by students as a secondary sexual characteristic but actually they are just a side-effect of the hormones that control puberty – make sure you don't get this muddled.

Genes and chromosomes

Your **genome** is one copy of all your genetic information (your DNA). Identical copies of your genome exist in all your **diploid** cells because they were produced by mitosis. Your **haploid** gametes (sperm or ova) were produced by meiosis and so only have half of your genome, because they would fuse during fertilisation and form a new diploid cell: a fertilised ovum.

Your genome consists of about 2 metres of DNA. This DNA fits inside the nucleus of most of your cells, although your red blood cells don't have any DNA to maximise the oxygen they can carry. To fit this length of DNA into your microscopic cells, it is arranged neatly into shapes called **chromosomes**. Your genome is made from 46 chromosomes. These come in 23 pairs, because half were present in each of the parent sperm and ova that made you. These chromosomes paired up during the fertilisation process.

Most other animals and plants have different numbers of chromosomes from humans (their 'diploid number'). For example, lettuces have nine pairs and mosquitos only have one pair.

Chromosomes have regions that contain the DNA code needed to make proteins (see page 82). These regions that contain the DNA code are called **genes**. Because you inherited one chromosome from each parent, you will have two copies of each gene. These copies are called **alleles**.

> **Genome:** One copy of all the DNA found in your diploid body cells.
>
> **Diploid:** Describes a cell or cell nucleus with two sets of chromosomes (one from each parent).
>
> **Haploid:** Describes a cell or nucleus of a gamete that has an unpaired set of chromosomes.
>
> **Chromosome:** A thread-like structure made from DNA which contains many genes.
>
> **Gene:** A section of a chromosome made from DNA that possesses the code needed to make a protein.
>
> **Alleles:** Two copies of the same gene, one from your mother and the other from your father.

We think there are about 24 000 genes that carry the instructions to make us. The rest of the DNA is **non-coding** – it does not make proteins and so we call this 'junk DNA'.

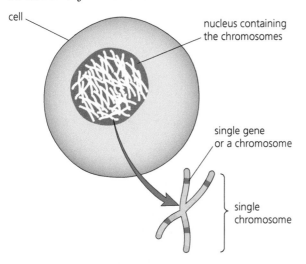

Non-coding: A description used to refer to a section of your genome that does not possess the instructions to make a protein. Scientists are still unsure about the purpose of these regions of our DNA.

Figure 5.9 **Make sure that you can identify the nucleus, chromosomes and the gene locations on the chromosomes.**

Revision activity

Copy Figure 5.9 without the labels and test yourself by adding them from memory. Check you are right by comparing your diagram with Figure 5.9.

The structure of DNA

REVISED

DNA is a polymer that is made from two strands that form a **double helix**. This structure is shown in Figure 5.10. The DNA strands are made from alternating sugar and phosphate molecules. Between these two strands are DNA bases, which are joined by weak hydrogen bonds to form **complementary base pairs**.

The base pairs that always pair are:
- adenine (A) and thymine (T)
- cytosine (C) and guanine (G).

Each base joined with its sugar and phosphate molecules is called a **nucleotide**. These are repeating units within the DNA polymer.

These same four bases make up the DNA of all life on Earth, so DNA is universal. You have over 4 billion DNA base pairs in your genome.

Double helix: The characteristic spiral structure of DNA.

Complementary base pairs: The four bases A–T and G–C, which are always paired together in DNA. They make up the 'rungs' in the double helix.

Nucleotide: A DNA base together with a sugar and a phosphate molecule, which makes up the backbone of the double helix.

Exam tip

You should be able to describe the structure of DNA and define the term 'genome' as 'one copy of all the DNA found in your diploid body cells', because it often comes up in the exam.

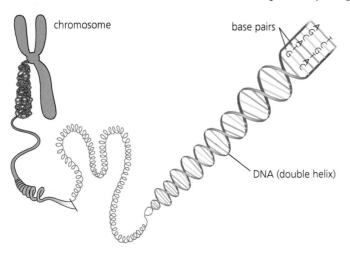

Figure 5.10 **DNA is coiled into a double helix and wound into chromosomes.**

Protein synthesis

Proteins are large, complex molecules that are made within our cells. Different proteins perform different functions, and some particularly important ones include:

- enzymes, which speed up all the reactions in your cells (see page 12)
- hormones, which are chemical messengers important in regulating growth and the menstrual cycle, for example (see page 77)
- antibodies, which are part of your immune response to infection (see page 49)
- those in your cell membranes that let substances in and out.

Many proteins only last for days in your body before they are broken down by your liver into urea, which is then excreted. This means that many of your cells must continuously carry out the process of making new proteins (protein synthesis).

Proteins are made from smaller molecules called amino acids. To make each protein, long chains of amino acids are folded into the correct shape. There are 20 different types of amino acid, which must be positioned in the correct sequence to make a protein. A protein molecule is often made from hundreds of these amino acids, as some types of amino acid will appear many times in a sequence (and others not at all).

The instructions on how to place these amino acids in the correct sequence are found within the DNA of a gene. In a gene's DNA, the sequence of the four DNA bases (adenine, thymine, cytosine and guanine) determines the order of amino acids to be placed in the protein. Three adjacent bases provides the code for a particular amino acid. These three bases are called a codon (for example AGA – adenine-guanine-adenine). There are 64 different possible combinations of bases that can make up a codon but only 20 different amino acids. Therefore, some different codons provide the instructions for the same amino acid. For example, the codons TCT, TCC, TCA and TCG all provide the same code for an amino acid called serine.

Protein synthesis is a two-step process. The first part is called transcription and the second is called translation.

Transcription

Transcription is the first part of protein synthesis and it occurs in the nucleus of a cell. In transcription, an enzyme attaches to a strand of DNA at the beginning of a gene. The enzyme then moves along the gene, breaking the hydrogen bonds between the DNA bases that make the double-stranded helix (almost like it is unzipping the DNA). As the enzyme moves along the gene it attracts messenger RNA (mRNA) bases to match one side of the DNA strand.

These mRNA bases match by making an opposite copy of the original DNA strand. These opposites are as follows:

- Opposite a G is a C
- Opposite a C is a G
- Opposite a T is an A
- Opposite an A is a U (uracil) base. All T (thymine) DNA bases are replaced by uracil (U) bases in RNA.

When the enzyme reaches the end of the gene it detaches and the mRNA strand is complete. This strand might include hundreds or thousands of

---

Protein synthesis: The formation of a protein by adding amino acids in the correct sequence. This occurs in ribosomes found in the cytoplasm of cells.

Codon: Three adjacent bases that code for an amino acid.

Transcription: The first part of protein synthesis in which an mRNA copy of a gene sequence of DNA is made.

Translation: The second part of protein synthesis in which amino acids are placed into the correct sequence by tRNA molecules in a ribosome.

Messenger RNA (mRNA) (ribonucleic acid): A chain of bases that makes a copy of a gene's DNA during transcription.

Typical mistake

Students often forget that uracil (U) replaces thymine (T) bases in RNA because they're thinking about opposites, so it's important to remember that thymine is the exception.

bases. The mRNA strand is now able to leave the nucleus and pass into the cytoplasm. (The chromosomes are too large to do this.) The hydrogen bonds between the DNA bases are then reformed.

Translation

Translation is the second and final part of protein synthesis. It begins when the end of a strand of mRNA attaches to a **ribosome** in the cytoplasm. The ribosome moves along the mRNA strand and causes the reactions that link the amino acids together.

In this process, for every three mRNA bases (a codon), the ribosome attracts a **transfer RNA (tRNA)** molecule. The three bases on this tRNA molecule are the opposite (the **anti-codon**) of the three mRNA bases in the codon. The tRNA molecule leaves behind its amino acid. The next tRNA molecule attaches to the next three mRNA bases, and then it too leaves behind its amino acid. This process is repeated hundreds of times, with each tRNA molecule bringing the correct amino acid to add to the growing chain.

> **Exam tip**
>
> A chain of amino acids is called a **polypeptide**. The polypeptide is folded into the correct shape to make a protein.

> **Ribosome**: A small cell organelle in the cytoplasm in which proteins are made.
>
> **Transfer RNA (tRNA) (ribonucleic acid)**: A molecule that lines up with every codon (three bases) of mRNA to add an amino acid and make a protein during translation.
>
> **Anti-codon**: Three adjacent nucleotides in a tRNA molecule.
>
> **Polypeptide**: A chain of amino acids that has yet to be folded into a protein.

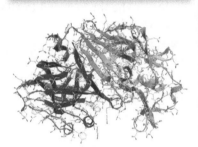

Figure 5.11 Polypeptides are chains of amino acids that are folded into proteins with complex shapes.

> **Now test yourself**   TESTED ☐
>
> 25 Define the term genome.
> 26 What are chromosomes?
> 27 How many chromosomes are in the diploid body cells of a human?
> 28 What is a gene?
> 29 What are the names of the four bases in DNA?
> 30 What links base pairs together in DNA?
> 31 What is a nucleotide?
> 32 Define the term transcription.
> 33 What happens during the translation step of protein synthesis?
> 34 What is the difference between a polypeptide and a protein?
>
> Answers on p. 127

> **Exam tip**
>
> You should be able to recall a simple description of protein synthesis, including how the structure of DNA affects the protein being made.

> **Revision activity**
>
> Write out the procedure for protein synthesis as a flow diagram to help you remember it.

Patterns of inheritance

Monohybrid inheritance   REVISED ☐

Sexual reproduction involves two parents that produce genetically different offspring. These offspring possess genes from both parents. Some characteristics (such as blood group) are controlled by a single gene and you inherit a gene for each of these characteristics from each parent. This means you have two copies of each gene, and we call these copies alleles.

> **Exam tip**
>
> Often exam questions ask you to define key terms. There are a lot of key terms in this section. Make sure you can define each one.

Eye colour

The sperm and the ovum that made you will each have had a gene for eye colour. These alleles will determine your eye colour depending on whether they are both the same gene or whether one gene is dominant. To look at gene characteristics we often use letters that we call genotypes. The phenotype is the physical characteristics of an organism, as determined by its genotype. An organism's phenotypic features are therefore controlled by their genes. Eye colour and blood group are examples of these features.

You inherited a gene from each of your parents, so there are two letters in a genotype. For genes that are dominant we use a capital letter; for genes that are recessive we use a lowercase letter. In the example of eyes, brown eyes are dominant over blue eyes so we would label the gene for brown eyes B. Blue eyes are recessive because they are a gene that can be dominated so we would label this gene b. Inheriting a B from both parents (BB) is called homozygous dominant. Inheriting a b from both parents (bb) is called homozygous recessive. Inheriting one of each (Bb) is called heterozygous.

Table 5.5 **The three possible allele combinations (genotypes) for brown/blue eye colour**

Genotype	Phenotype	Terminology
BB	Brown eyes	Homozygous dominant
bb	Blue eyes	Homozygous recessive
Bb	Brown eyes	Heterozygous

We complete genetic crosses in Punnett squares to see the likelihood of inheriting certain characteristics. In the example in Figure 5.12 the mother's genotype is BB, which means her phenotype is brown eyes. Because she is homozygous dominant, all her ova will have B genes. The father's genotype is bb. Because he is homozygous recessive all his sperm will have b genes. That means that each of the four possible combinations for their children are the same: Bb. As a result, all their children must be heterozygous and have brown eyes.

The outcomes of these genetic crosses can also be given as percentages or ratios. So, if we can predict that a quarter of the possible genotypes will give blue eyes and three-quarters will give brown, we would say that there is a 25% chance of blue eyes and a 75% chance of brown. We could also express these percentages as the following ratio – 1:3.

The probabilities of other characteristics, such as whether you have attached ear lobes or can roll your tongue, can also be determined using Punnett squares. Both the phenotypes and genotypes of these characteristics are shown in Table 5.6.

Table 5.6 **The three possible genotypes and phenotypes for ear lobes and tongue rolling**

Terminology	Ear genotype	Ear phenotype	Tongue genotype	Tongue phenotype
Homozygous dominant	EE	Free lobes	TT	Can roll
Homozygous recessive	ee	Attached lobes	tt	Can't roll
Heterozygous	Ee	Free lobes	Tt	Can roll

Dominant: A characteristic that will display if inherited from one or both parents.

Genotype: The genetic make-up of an organism, which is often represented by letters.

Phenotype: The physical characteristics of an organism.

Recessive: A characteristic that will display only if inherited from both parents.

Homozygous dominant: A genotype with two dominant alleles.

Homozygous recessive: A genotype with two recessive alleles.

Heterozygous: A genotype with one dominant and one recessive allele.

Punnett square: A grid that makes determining the chance of inheriting a characteristic easier to understand.

Revision activity

Draw out Table 5.5 with only the headings. Try to fill in the rest of the table from memory to help you revise.

Exam tip

Be careful when filling in Punnett squares that your capital and lowercase letters look different. 'C' and 'c' (for example) can look very similar.

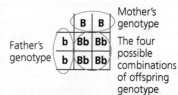

Figure 5.12 **The four possible genotypic offspring of the parents whose genotypes were BB and bb**

Exam practice answers and quick quizzes at **www.hoddereducation.co.uk/myrevisionnotesdownloads**

Most characteristics are a result of multiple genes interacting, rather than a single gene as described above.

> **Exam tip**
>
> You may be asked to explain genetic crosses using direct proportions and simple ratios. It is important you work in a logical way when completing these Punnett squares and their analysis because otherwise you could make a simple mistake.

Polygenic inheritance

REVISED

We now know that the features of most organisms (their phenotypes) are controlled by the interaction of several or many genes together, not individual genes as described in the monohybrid model of inheritance above. This is called polygenic inheritance (polygenic means 'many genes'). Height is an example of polygenic inheritance, as several hundred genes (and environmental factors such as diet) control someone's height.

Codominance

REVISED

In some situations, two or more alleles are expressed to give a phenotype (a physical characteristic). This is called **codominance** and it is denoted by two capital letters. One such example is the colour of feathers in chickens:

- The allele for white feathers is W.
- The allele for black feathers is B.
- This means that WW is white, BB is black, but crucially WB (or BW) is speckled (black and white feathers).

> Codominance: When the alleles of a gene pair are both fully expressed, resulting in offspring with a phenotype that is neither dominant nor recessive.

Blood groups in humans is another example of codominance. In this example three alleles are involved: A, B and O.

- A and B are equally dominant and both dominate O.
- Alleles AA and AO give blood group A.
- BB and BO give blood group B.
- A and B together give blood group AB.
- Finally, alleles O and O give blood group O.

Inherited disorders and family pedigrees

REVISED

Communicable diseases are caused by pathogens but inherited conditions, like cystic fibrosis, are called disorders. The people who inherit these conditions from their parents are often called sufferers.

About one in every 10 000 people in the UK has cystic fibrosis. This is a disorder of cell membranes where sufferers produce unusually sticky mucus in their lungs, digestive system and reproductive system. This mucus often becomes infected. Although frequent physiotherapy sessions may help to remove the mucus, sadly there is no cure and sufferers have a reduced life expectancy.

Sufferers of cystic fibrosis inherit a recessive allele from both parents and therefore all sufferers are homozygous recessive. If sufferers had inherited one dominant gene from either parent, this would make them heterozygous and would mean they didn't have cystic fibrosis. However, people who are heterozygous for the disorder are able to pass it on to their children, so in this case we call them carriers for the disorder.

A family tree can show the inheritance of characteristics over multiple generations. Every generation has its own horizontal line, with the oldest family members at the top.

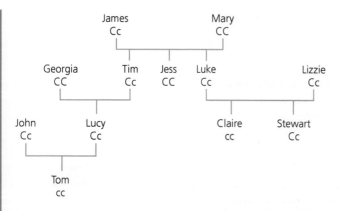

Figure 5.13 The gene for cystic fibrosis is shown by the letter 'c' in this family tree. CC is a person not suffering from cystic fibrosis. Cc is a carrier who doesn't have the disorder. cc is a sufferer with cystic fibrosis.

Inheritance of sex

We call the 23rd pair of chromosomes in humans the sex chromosomes, because this is what determines whether someone is male or female. As with all genetic characteristics, we use letters to describe them:

- All ova are X.
- Approximately half of sperm are also X, but the other half are Y.
- An X ovum and an X sperm develop into a female – XX.
- An X ovum and a Y sperm develop into a male – XY.

Exam tip

Exam questions may ask you to carry out a genetic cross to show sex inheritance. You should be able to use direct proportions and simple ratios to analyse these results. It is also important you can use these results to explain why approximately half the population is female and half is male.

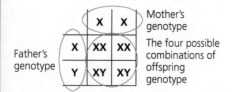

Figure 5.14 The relative proportions of male and female offspring resulting from this cross explain why approximately 50% of the human population is of each sex.

Mitosis

Your diploid body cells are continually dying. For example, it is unlikely you have any blood cells alive in your body from a year ago – they will all have been replaced at least once. As your cells are continually dying, this means that they need replacing. **Mitosis** and the cell cycle explain how this replacement process occurs. Mitosis exchanges damaged or old cells with identical replacements. Mistakes in this process could lead to cancer. This medical condition occurs when cells divide uncontrollably to form a lump (tumour).

Mitosis: The formation of identical diploid cells for growth, repair, cloning or asexual reproduction.

Without mitosis, your body would not have developed from the one diploid cell that was originally formed when your father's sperm fertilised your mother's ovum. Additionally, without the ability to replace cells in this way, any cuts, burns or other damage to your body could not heal. Mitosis is essential for growth, development and repair in multicellular organisms.

During the cell cycle, all chromosomes are copied. This doubles the number of chromosomes in humans from 23 pairs to 46 pairs, or 92 in total. The number of cellular components, such as ribosomes and mitochondria, also doubles. The cell finally divides along its length into two identical 'daughter' cells, each with an entire copy of the organism's genome.

Steps in mitosis

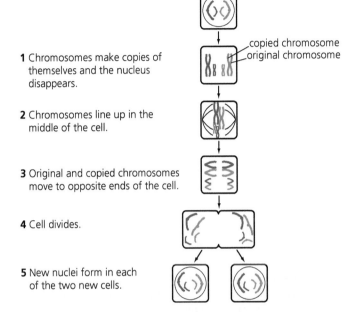

1 Chromosomes make copies of themselves and the nucleus disappears.

copied chromosome
original chromosome

2 Chromosomes line up in the middle of the cell.

3 Original and copied chromosomes move to opposite ends of the cell.

4 Cell divides.

5 New nuclei form in each of the two new cells.

Figure 5.15 **The main steps in mitosis for a cell with just two pairs of chromosomes. Note that the two new cells at the bottom of the diagram are identical to each other and also to the original 'parent' cell at the top in terms of chromosome numbers.**

> **Revision activity**
>
> Write out the procedure for mitosis as a flow diagram to help you remember it.

Meiosis

REVISED

Gametes are sex cells that are produced in meiosis. These cells are haploid and so have half the number of chromosomes of a normal diploid body cell. A diploid human body cell, like a nerve or muscle cell, has 46 chromosomes, or 23 pairs. Haploid human sperm and ova therefore have 23 chromosomes. During fertilisation, the sperm and ovum fuse together to make a genetically new, diploid fertilised ovum. Fertilisation is a random process and so produces genetically different offspring. At fertilisation, the two sets of 23 chromosomes fuse to form a new diploid cell with 23 pairs of chromosomes. This new diploid cell divides by mitosis and it can then differentiate for different functions as it grows into an adult organism.

> **Exam tip**
>
> You should be able to state that meiosis leads to non-identical cells being formed, while mitosis leads to identical cells being formed.

> **Exam tip**
>
> To help remember the difference between meiosis and mitosis, it is helpful to note that the first letters of meiosis spell out Making Eggs In Ovaries, Sperm In Scrotum.

The process of meiosis

The steps in the process of meiosis are shown in Figure 5.16.

1 Chromosomes make copies of themselves and the nucleus disappears.

2 Chromosome pairs line up and swap pieces of information (DNA crossover).

3 Cell divides.

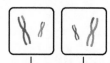

4 Chromosomes line up.

5 Original and copied chromosomes move to opposite ends of the cell. Cell divides for a second time.

6 Four new nuclei form.

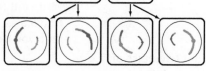

Figure 5.16 **The main stages in meiosis for a cell with just two pairs of chromosomes. Once the matching chromosomes have been paired up, they may swap pieces of DNA between them. This is called DNA crossover. The cell divides twice and ends up as four daughter cells (gametes), each with half the original number of chromosomes.**

During meiosis:

- Copies of the organism's DNA are made (a human cell would now have 92 chromosomes).
- DNA is then exchanged to ensure all gametes are genetically different from each other.
- The cell then divides on two separate occasions (firstly back to two diploid cells with 46 chromosomes and then to four haploid cells with 23 chromosomes).

Table 5.7 **The key similarities and differences between mitosis and meiosis**

Feature	Mitosis	Meiosis
Number of cells at beginning	One	One
Type of cell at beginning	Diploid body cell (23 pairs of chromosomes)	Diploid body cell (23 pairs of chromosomes)
Number of cells at end (daughter cells)	Two	Four
Type of cells at end	Diploid body cells (23 pairs of chromosomes)	Haploid body cells (23 chromosomes)
Number of divisions	One	Two
Identical or non-identical cells produced	Identical	Non-identical
Purpose	Growth and repair Asexual reproduction	Producing gametes
Location	Everywhere	Sex organs only

Now test yourself

TESTED

35 Define the term alleles.
36 What is the difference between genotype and phenotype?
37 Define the terms homozygous dominant, homozygous recessive and heterozygous.
38 What is the probability (as a percentage) that a child will be homozygous recessive for an inherited disorder if both parents are carriers?
39 What does polygenic inheritance mean?
40 What genotype results in a male human baby?
41 What are the products of mitosis?
42 What is the purpose of mitosis?
43 What are the products of meiosis?
44 What is the purpose of meiosis?

Answers on p. 127

Variation, change and evolution

Variation

REVISED

Variation is the sum of all the differences between two organisms of the same or different species. In other words it can mean all the differences between two cats (the same species) or all the differences between a cat and a dog.

Causes of variation

Variation can be caused by:
- environmental factors only, which often affect physical appearance (such as accidents that cause scars)
- genetic factors only, which control things like blood group and eye colour
- both environmental and genetic factors combined, which affect most features (such as your weight and height).

When genetic and environmental factors work together, a person's genome interacts with the environment to influence their development.

Types of variation

There are two main types of variation: continuous and discontinuous. Continuous data have values across a range, anywhere between two extreme values. Height is an example of continuous data; you could, for example, have 140 cm and 190 cm as your two extremes and your data may then fall anywhere between these two points. Continuous data are usually presented in a line graph with a line of best fit.

Discontinuous data come in discrete groups. Blood group is an example of a characteristic that shows discontinuous variation, as you can only be A, B, AB or O; data cannot be in between any of these types. Discontinuous data are usually presented in a bar chart.

Variation: The differences that exist within a species or between different species.

Species: The smallest group of classifying organisms. All members of a species are able to interbreed to produce fertile offspring.

Continuous (data): Data that come in a range and not in groups.

Discontinuous (data): Data that come in groups and not a range.

Normally distributed variation

Line graphs of continuous data often show a characteristic 'bell-shaped graph'. This is shown in Figure 5.17. We call this a **normal distribution** and it means most values are towards the middle, with an ever-smaller number of values towards the outer limits.

Normal distribution: Data that are more common around a mean and that form a bell-shaped graph.

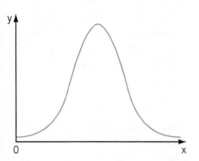

Figure 5.17 **Bell-shaped graphs show normally distributed data. The most common values are in the middle and the least common values are at both ends.**

Mutation

Changes to the sequence of DNA are called **mutations**. Mutations are rare and random, but can also be inherited. Mutations can happen naturally or they can be caused by things like **carcinogenic** chemicals. Examples of these chemicals include tar and others found in tobacco, as well as asbestos, which was a building material used due to its resistance to fire and insulating qualities. Cutting or drilling asbestos produces fibres that can cause cancer of the respiratory system when inhaled.

Mutations can also be caused by **ionising radiation**. Examples of this type of radiation include X-rays, alpha and beta particles and gamma rays.

Mutation: A permanent change to DNA, which may be advantageous, disadvantageous or have no effect.

Carcinogen: A cancer-causing substance.

Ionising radiation: Radiation such as UV rays, X-rays and gamma rays that can cause mutations to DNA.

The effects of mutations

Mutations often have no effect on the organism but they can also be either advantageous or disadvantageous. If a mutation happens in part of the genome that does not code for a gene, then it is unlikely to have an effect. Very rarely a mutation that causes an advantageous change in an organism's phenotype could spread through a whole population. This could then lead to evolutionary change.

Mutations can involve swapping the order of bases in a sequence, or deleting them or adding them. If a mutation occurs in a gene, the sequence of DNA bases, and so the sequence of amino acids, might be changed. This mutation could give the organism an advantage by making a better protein, or a disadvantage by not making a protein at all or by making a less effective one. It could therefore change the phenotype of the organism. Following a mutation, an enzyme may no longer fit the substrate binding site or a structural protein may lose its strength. However, most mutations do not alter the shape of the protein and so have no effect on organisms.

Typical mistake

A common mistake in exams is to write that all mutations have negative consequences. It is important to remember that many have little or no effect on an organism at all and some have positive consequences.

Evolution

REVISED

Evolution explains how the millions of different species alive today developed from one **common ancestor**. It explains how over many generations, tiny changes in individuals gave them an advantage and allowed them to become better suited to their surroundings. Eventually these small differences added up to make different species.

Charles Darwin and the theory of evolution by natural selection

Charles Darwin (1809-82) is considered the father of evolution. In his youth, he spent several years travelling around the world on a ship called *HMS Beagle*. During this voyage, he stopped at the Galapagos Islands, off the coast of Ecuador, where he collected animal specimens and made extensive observations. Darwin's background knowledge of geology and fossils, combined with his Galapagos observations (particularly of finches), helped him to develop his theory of evolution by natural selection. This theory says:

● Individual organisms within a particular species show variation in their characteristics.
● Those individuals with characteristics most suited to their environment are most likely to survive and breed.
● These advantageous characteristics are therefore more likely to be passed on to the next generation.

Darwin was originally reluctant to publish his findings because he was worried about the reaction of the Church, which was very powerful at this time. When he eventually did publish, Darwin's fears were well-founded. His book *On the Origin of Species* (1859) proved very controversial for the following reasons:

● The theory challenged the idea that God made all life on Earth.
● There was insufficient evidence at the time to convince many scientists.
● The mechanism of genetic inheritance was only discovered 50 years after publication.

Speciation

Alfred Russel Wallace (1823–1913) was a naturalist who was inspired by Darwin's work. Wallace also voyaged around the world and spent eight years studying the animals and plants of what we now call Singapore, Malaysia and Indonesia. There is a narrow strait of water between these regions and Wallace noticed key differences in the species found on either side of this line. This region is now called the Wallace line.

From these observations, Wallace developed his own theory of evolution. Amazingly, Wallace sent his theory to Darwin to get his opinion. At this point, Darwin had not yet published his work so both scientists agreed to

Evolution: The theory first proposed by Charles Darwin that the different species found today developed as a result of the accumulation of small advantages that were then passed through generations.

Common ancestor: A single organism from which others have evolved.

Figure 5.18 Darwin's finches as drawn by Darwin himself

Exam tip

You should learn the definition of evolution as 'a change in the inherited characteristics of a population over time through a process of natural selection, which may result in the formation of new species'. It is likely you will need it in the exam.

Revision activity

Write out the procedure for evolution as a flow diagram to help you remember it.

jointly publish what they had discovered. Despite their joint publication, Darwin is most well-remembered for the theory of evolution. Wallace is better remembered for his work on **speciation**. Speciation is the development of different species because of evolution.

Speciation: The process of forming new species of life as a result of evolution.

Evidence for evolution

Since Darwin published his theory, evidence for the process of evolution has developed to the point where most scientists agree with it. Most criticism of the theory now comes from religious groups who believe in **creationism**. This is the belief that God created the universe and all life in it.

Fossils

Fossils are the remains of dead organisms that have been preserved in rock for millions of years. Fossils of whole organisms or parts of organisms can be preserved when their tissues are slowly replaced by minerals as they decay. Other examples of fossils are traces of organisms that might include dinosaur footprints, burrows or eggs. In special places like **peat** bogs, conditions are such that the rate of decay is very slow, which preserves fossils especially well.

The **fossil record** is the information provided by all the fossils that have ever been discovered. This record can show us how significant the changes to species have been over time. However, there are gaps in the fossil record because not all fossils have been found and not all organisms leave fossils behind. These gaps can be explained by phenomena such as destruction by magma and also by the fact that not all parts of soft-bodied organisms can become fossils.

Creationism: A belief that God created all the organisms on Earth rather than them evolving from a common ancestor.

Peat: Partially decayed vegetation.

Fossil record: A record of all of the fossils that have been discovered so far.

Antibiotic: A medicine, for example penicillin, that slows or stops the growth of harmful bacteria.

Methicillin-resistant *Staphylococcus aureus* (MRSA): A bacterium that has evolved resistance to antibiotics.

Antibiotic-resistant bacteria

Sir Alexander Fleming (1881–1955) discovered the first **antibiotic** called penicillin in 1928. Since then scientists have developed several other antibiotics. However, it seems that bacteria are evolving to become immune to our antibiotics faster than we can develop new ones.

A common strain of bacteria that has developed resistance to antibiotics is **MRSA (methicillin-resistant** *Staphylococcus aureus*). MRSA is a communicable pathogen that kills several hundred people per year in the UK. It causes a fever and aches and pains. It also causes red, swollen bumps on the skin that are painful to the touch.

To reduce the speed at which antibiotic resistance develops, it is important that we reduce the use of antibiotics generally, and always finish the full course of medicine we are given.

Bacteria reproduce much more quickly than most animals. This has provided evidence for evolution, because we have seen bacteria evolve resistance to antibiotics in our lifetime. This speed of reproduction also causes problems because without an effective way of slowing it (such as antibiotics), the population of harmful bacteria can increase very quickly, meaning the risk of others becoming infected is hugely increased. Once infected, the speed of bacterial reproduction can cause the patient to have severe symptoms even before diagnosis can occur and treatment begin.

Now test yourself

TESTED ☐

45 Describe the causes of variation.
46 State two examples of environmental variation.
47 Define the term continuous data.
48 What type of distribution is observed in a bell-shaped graph?
49 Give three ways that a mutation can change the sequence of DNA bases.
50 Define the term common ancestor.
51 Describe Darwin's theory of evolution.
52 What are fossils and how do they provide evidence for evolution?
53 What is the significance of MRSA bacteria for the theory of evolution?
54 Why is it easier to study evolution in bacteria than in humans?

Answers on p. 127

Revision activity

Draw a mind map of all the content in this chapter. You could use the topic headings: reproduction in living organisms, reproduction in flowering plants, reproduction in humans, genes and chromosomes, patterns of inheritance, and variation, change and evolution.

Summary

- Sexual reproduction involves the fusion of a male and a female gamete to give the normal number of chromosomes.
- Asexual reproduction requires one parent (not two) and so there is no fusion of gametes. Offspring are genetically identical clones of the parent. Only mitosis, not meiosis, is involved.
- The male part of a flower is the stamen, made from the anther and filament. Pollen is contained in the anther. The female part of the flower is the carpel, made from the stigma, style and ovary.
- The female reproductive system comprises the ovaries, oviducts (fallopian tubes), uterus, cervix and vagina. The male reproductive system comprises the penis, scrotum, testes, urethra, vas deferens and prostate gland.
- The menstrual cycle is an approximately 28-day reproductive cycle in women that begins at puberty and continues to the menopause.
- A genome is all of an organism's genetic material (DNA). In eukaryotes, DNA is arranged into chromosomes. A gene is a section of DNA that codes for a sequence of amino acids to make a specific protein.
- DNA is made from four complementary bases, which are found in the following pairs: A with T and C with G.
- Protein synthesis is the manufacture of proteins.
- Some characteristics are controlled by one gene. However, most characteristics are controlled by the interaction of more than one gene.
- Alleles are pairs of genes, one inherited from each parent. The alleles present are an organism's genotype. Their expression is an organism's phenotype.

- Some characteristics, like chicken feather colour and human blood group, are controlled by more than one allele. This gives rise to codominance.
- Mitosis is a type of cell division. It produces two identical diploid cells for growth and repair.
- Meiosis is another type of cell division. Unlike mitosis, it produces four non-identical cells with half the DNA of the original cell. These are gametes (sex cells), which are sperm and ova in animals, and pollen and ova in flowering plants.
- Variation is the sum of all the differences between organisms of the same or different species. The causes of variation are genetic, environmental or combinations of both.
- Mutations in DNA can affect the order of amino acids and so the shape of a protein. Most mutations have no effect on the phenotype, although a few affect phenotype. Advantageous mutations can lead to changes in species.
- Evolution is a change in the inherited characteristics of a population over time through the process of natural selection, which may result in the formation of a new species.
- Darwin's theory of evolution by natural selection states that all species of life evolved from the same simple organisms over the last three billion years.
- Darwin's theory says that individual organisms within a species show a wide range of characteristics. Individuals with characteristics most suited to their environment are more likely to survive and breed. Their offspring are likely to inherit these characteristics.

Exam practice

1 Variation is a key part of the process of evolution.
   (a) Variation in eye colour and the ability to roll your tongue are examples of which type of variation? [1]
      A Continuous
      B Genetic
      C Genetic and environmental
      D Environmental
   (b) The data in the table show the different blood groups found in a survey of people.
      Draw a graph of the results to show these data. [6]

Blood group	A	B	AB	O
Number of people	21	5	2	24

   (c) Variation occurs, in part, because of a type of cell division.
      Describe the differences in the cells produced in mitosis and meiosis. [4]

2 Charles Darwin is thought of as the father of evolution.

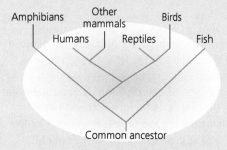

   (a) What does this image tell you about the evolutionary relationships between the five classes of vertebrate? [4]
   (b) Doctors are now prescribing fewer antibiotics in order to reduce the evolution of antibiotic-resistant bacteria. Describe the process of evolution, using antibiotic-resistant bacteria as an example. [5]

3 Inheritance is the passing of characteristics from one generation to the next in our DNA.
   (a) Name the missing parts A–D of the diagram on the right. [4]
   (b) State what sort of bonds hold the DNA base pairs together. [1]
   (c) Eye colour is one characteristic that is passed down. Copy and complete this Punnett square for eye colour. [2]

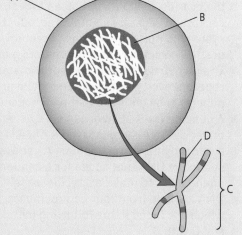

		Female alleles	
		B	b
Male alleles	B		
	B		

   (d) A male heterozygous for eye colour marries a homozygous recessive female. Copy and complete a Punnett square to show their possible offspring. Analyse your results to give percentage probabilities. [6]

	Female alleles	
Male alleles		

Answers and quick quizzes online

ONLINE

6 Ecology and the environment

The organism in the environment

Key terms

A **population** is the total number of all the organisms of the same species in a geographical area. This area could either be the whole planet or a much smaller part of it. A **community** is a group of two or more populations of different species in the same geographical area.

A **habitat** is the natural environment of a living organism. For example, the habitat of a lion is the plains of Africa. An **ecosystem** is a community of living organisms and the non-living parts of their environment. To survive and reproduce, all organisms need to take advantage of the resources from their surroundings as well as the other organisms that live with them. For example, lions must take advantage of available water in the plains and feed on animals lower in the food chain.

> **Population:** The total number of all the organisms of the same species, or the same group of species, that live in a particular geographical area.
>
> **Community:** A group of two or more populations of different species that live at the same time in the same geographical area.

Interdependence

All organisms within a community depend upon each other. Examples of this dependence might include food, pollination and seed dispersal. The organisms within each community will have evolved to have this dependence on each other and we call this sort of relationship **interdependence**.

A community in which high levels of interdependence are found is called 'stable'. In a stable community there is a balance between predators and prey. The numbers of predators and prey may rise and fall (as shown in Figure 6.1) but this interdependence means things will never get to the point where one organism kills off or totally outcompetes another. Because of this, the removal of one species from an ecosystem can affect all other organisms within it, as a whole area of the dependence will also be removed.

> **Habitat:** The area or natural environment of a living organism.
>
> **Ecosystem:** A community of different living organisms and their environment.
>
> **Interdependence:** A situation where all the organisms in a community depend upon each other.

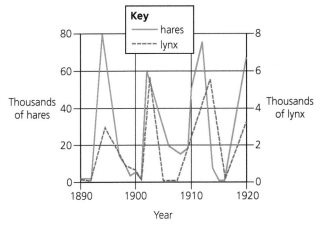

Figure 6.1 **Predator-prey cycling in Canadian lynx and snowshoe hare populations**

> **Exam tip**
>
> Exam questions will often ask you to explain the importance of interdependence in a community. Remember that interdependence is necessary for a healthy ecosystem. When interdependence is reduced, perhaps as a result of human activity, an ecosystem may fail.

Sampling

Sampling is the process of studying a small part of an ecosystem and drawing conclusions about a whole population from this sample. This process saves important time and money, as studying a whole population would be very difficult.

Quadrats

Quadrats are squares of wire that are often $0.25\,m^2$. They may be used during sampling to record the number of organisms in a specific area. Quadrats are used to count what is located within them in three main ways:

1 the number of individuals of a single species
2 the number of different species (as a measure of biodiversity)
3 the percentage cover of a particular species (such as grass).

There are two main ways in which you place quadrats depending upon what it is you are trying to investigate. If you want to know the numbers of a species in an area, or to compare two or more areas, you would place your quadrats randomly. If you want to investigate the change across a habitat, you would place your quadrats systematically along a line called a **transect**.

In all cases of sampling, it is important that you record multiple results. It is best to place at least 20 quadrats and calculate mean values from your results.

Random sampling using quadrats

If you are conducting random sampling, you should not simply stand in the middle of the habitat you wish to sample and throw your quadrat out randomly. This method would mean that the quadrat never landed at your feet or beyond the maximum distance of your throwing range and would therefore not be truly random. It is important that the placement is totally random in order to avoid bias in your results. The method you should ideally use is described in Required practical 1 on page 97.

Systematic sampling using quadrats

Systematic sampling looks for changes in the distribution of organisms as a result of changes within a habitat. Because of this, other abiotic factors like light intensity or moisture levels are often recorded alongside biotic ones.

If you wanted to see if or how the number and species of seaweed changed as you walked down a seashore, you would use systematic sampling. (This is different from estimating the total number of seaweed on the shore, for which you could use random sampling.) To do systematic sampling you would draw an imaginary line called a transect down your habitat – in this case the seashore. You would then place your quadrat at equal and regular distances down the transect – for example, every 10 metres. You would record the number and species of seaweed in each quadrat. At the same time, you would also need to record any abiotic factors that might help explain the changes you observe. In the case of the seashore, the position of seaweed is mainly determined by the number of hours a species of seaweed can be out of water. Those at the top of the shore have evolved to be out of the water for longer.

> **Sampling**: The process of studying a smaller amount of information in order to make wider conclusions.
>
> **Quadrat**: A square frame used in biological sampling.
>
> **Transect**: A line along which systematic sampling occurs.

> **Systematic sampling**: The regular distribution (that is, not random) of a survey to answer a specific question, usually about a trend.

Required practical 1

Investigate the population size of an organism in two different areas using quadrats

Equipment:
● Quadrat
● Ruler

Method:
This method is an example of random sampling:
● A starting location in one corner of the first area being studied was chosen.
● Random numbers were generated using a table or calculator.
● The first two random numbers were used as coordinates to place the first quadrat.
● Random numbers larger than the area were ignored.
● The total number of individuals of the species being investigated in this quadrat was counted.
● The experiment was then repeated many times using different random numbers as coordinates.
● The experiment was then repeated in its entirety in the second area.
● The results of the two experiments were then compared.

Results:
● A mean value of the number of organisms was calculated for each area.
● The total number of organisms in each area was estimated by multiplying the mean number of organisms per quadrat by the difference in size between the quadrat and the whole area.

Sampling small animals

Quadrats are ideal for sampling plants or very slow-moving animals like snails. However, faster-moving animals are more difficult to sample with this method. Table 6.1 shows three sampling techniques for small animals. Figure 6.2 (on the following page) shows how a pooter can be used to collect even smaller animals.

Exam tip

If an exam question asks you about the principles of sampling, it is important that you write that often sampling is random and should include a sufficiently high number of samples (and not just three repetitions).

Table 6.1 **Three methods for sampling small animals**

	Sweep net	Pond net	Pitfall trap
Sampling type			
Use	Sampling small animals from vegetation	Collecting small animals from water	Collecting small, crawling land animals
Method	The flat lower section of the net is dragged along the vegetation.	When kick-sampling, the net is placed on the river bed downstream and further upstream the bed is gently kicked. Any organisms dislodged are gently washed into the net.	A small amount of food may be placed in the trap to attract animals.

Revision activity

Draw out Table 6.1 with only the headings and the first column completed.
Try to fill in the rest of the table from memory to help you to revise.

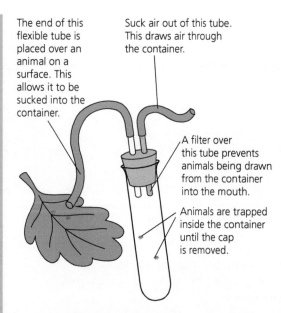

The end of this flexible tube is placed over an animal on a surface. This allows it to be sucked into the container.

Suck air out of this tube. This draws air through the container.

A filter over this tube prevents animals being drawn from the container into the mouth.

Animals are trapped inside the container until the cap is removed.

Figure 6.2 **Small animals, like insects, can be sucked up into a pooter.**

Biodiversity

REVISED

Biodiversity is a measure of the variety of all the different species of organisms on Earth, or within a particular ecosystem. Areas of low biodiversity include deserts and the polar regions, as there are fewer organisms in these areas. Areas of high biodiversity include tropical rainforests and ancient oak woodlands where many organisms will thrive.

A relatively high biodiversity in an ecosystem helps to ensure its stability by reducing the dependence of one species on another, whether that's for food, shelter or the maintenance of the physical environment. Crucially, the future of the human species on Earth depends upon us maintaining a good level of biodiversity across the planet. Many of our activities, including **deforestation**, pollution and burning of fossil fuels, reduce biodiversity and therefore cause us problems. The effects of deforestation are described in detail on page 108.

> **Typical mistake**
>
> Don't forget that the term biodiversity can be used for both the variety of life on Earth, and the variety within a specific ecosystem. Some people think that the term is only used to refer to all of life on Earth.

Conservation is one way in which biodiversity can be maintained. This approach involves protecting certain areas with high ecological importance like peat bogs, ancient forests and marine regions like coral reefs. Other ways of maintaining biodiversity include:

- zoo **breeding programmes** to increase the number of endangered species
- reintroduction of hedgerows in areas that have been **intensively farmed**, because these provide habitats for far more species than open fields do
- reducing deforestation and carbon dioxide emissions
- reducing waste, reusing and recycling rather than dumping in landfills.

Biodiversity: A measure of the variety of all the different species of organisms on Earth, or within a particular ecosystem.

Deforestation: The cutting down of trees (often on a large scale).

Conservation: Protecting an ecosystem or species of organism from declining numbers and often extinction.

Breeding programme: An activity that zoos undertake which involves breeding captive animals that may otherwise be endangered in order to increase their numbers and so the gene pool.

Intensive farming: Industrial agriculture to maximise yield, often involving the use of machines, chemical fertilisers and pesticides, and the removal of hedgerows to make bigger fields.

Required practical 2

Measure biodiversity using quadrats

Equipment:
- Quadrat
- Ruler

Method:

This method is an example of random sampling:

- A starting location in one corner of the ecosystem was selected.
- Random numbers were generated using a table or calculator.
- The first two random numbers were used as coordinates to place the first quadrat.
- Random numbers that were larger than the area were ignored.
- The total number of different species in this quadrat was recorded.
- This provided a measure of biodiversity.
- The experiment was then repeated many times using different random numbers.

Results:
- A mean value per quadrat was calculated.
- Areas with a higher mean had a greater biodiversity.

Biotic factors

REVISED

Biotic factors are the living parts of the environment. Examples of important biotic factors include the availability of food, numbers of predators, introduction of a pathogen that causes a communicable disease, and competition between species. Within an ecosystem, more organisms will be found where food is available and where there are fewer predators. The number of organisms in an area is limited by interactions between prey and predator and also competition over resources. The introduction of a new pathogen can also drastically reduce the number of organisms across an entire ecosystem.

Changes in biotic factors often result from newly introduced species. If these new species damage the local ecosystem by outcompeting existing species, they are called invasive species. The cane toad in Australia and the grey squirrel in the UK are invasive species.

Biotic factors: The living parts of the environment.

Invasive species: An organism that is not native and causes negative effects on an ecosystem.

Abiotic factors

REVISED

Abiotic factors are the non-living parts of the environment. These factors can be chemical or physical, but not biological. Examples of some of the important abiotic factors for plants are:

- Light intensity – Plants grow at light intensities that suit them. Most plants do better in bright light because they can photosynthesise more effectively, but not all plants grow best in brightness. Some plants, like ivy, grow well in the shade.

Abiotic factors: The non-living parts of an environment.

- Temperature – Temperature affects the distribution of plants. For example, a plant like the tropical orchid that is adapted to grow in hot, humid conditions like the rainforest will not survive outside in colder climates.
- Moisture levels – Moisture levels drastically affect the growth of plants. For example, water plants like water lilies cannot grow on dry land and plants like cacti that are used to dry conditions cannot grow in the wet.
- Mineral content – Many plants can only grow well in rich, fertile soil. However, carnivorous plants like the Venus flytrap and pitcher plants have evolved the ability to catch insects because they grow in nutrient-poor soils. The nutrients provided by the decaying insects make up for those they cannot absorb through their roots.

> **Exam tip**
>
> Exam questions about abiotic factors such as light intensity, temperature and water content of soil will often ask you to explain how these factors affect the distribution of organisms. Make sure you can write a sentence about each of these in detail.

Now test yourself

TESTED

1 Define the term population.
2 Define the term community.
3 What is interdependence?
4 Why do scientists sample an area rather than recording results for the whole area?
5 Define the term sampling.
6 State three ways in which quadrats are used.
7 When would a transect be used instead of random sampling?
8 Define the term biodiversity.
9 State two examples of abiotic factors that could affect the distribution of plants in an environment.
10 State two examples of biotic factors that could affect the distribution of animals in an environment.

Answers on p. 128

> **Typical mistake**
>
> If an exam question asks about how to carry out sampling in order to relate the distribution of organisms to an abiotic factor such as amount of water in the soil, don't forget you will need to use systematic sampling along a transect in your answer. If the question asks you to compare two areas, then you will need to use random sampling of quadrats in your answer.

Feeding relationships

Trophic levels of food chains and webs

REVISED

Food chains show the feeding relationships between organisms in an ecosystem.

grass ⟶ cow ⟶ human

All the food chains of an area can be compiled into a food web. In the food chain above, humans eat cows, while the cows eat grass. The arrows show the flow of energy from one organism to another. A **trophic level** is any level in a food chain. Levels in a food chain almost always follow this pattern:

- The first trophic level is plants or algae, which make their own food by photosynthesis. These are called **producers**.
- The second trophic level is herbivores, which eat plants or algae. These are called primary **consumers**.
- The third trophic level is carnivores that eat herbivores. These are called secondary consumers.
- The fourth trophic level is carnivores that eat other carnivores. These are called tertiary consumers.

There may be additional levels beyond those listed above, but food chains rarely exceed six trophic levels.

> **Trophic level:** A stage in a feeding relationship, which represents an organism in a food chain or a group of organisms in a food web.
>
> **Producer:** Any organism that photosynthesises (a plant or algae).
>
> **Consumer:** Any organism in a feeding relationship that eats other organisms for food.

Other key terms you should know in relation to food chains include **apex predators** and decomposers. Apex predators are the carnivores at the tops of food chains, which have no predators that prey on them. Decomposers are organisms that break down dead plant and animal matter by secreting enzymes into the environment. They play a very important role in recycling nutrients in an ecosystem.

Producers

Producers are the photosynthetic plants and algae found at the lowest trophic level of almost all food chains. During photosynthesis, these species turn carbon dioxide and water into glucose and oxygen using sunlight. The glucose that is generated then supports all life at higher trophic levels. Producers are called **autotrophs** because they 'feed' themselves.

Consumers

Consumers (also known as **heterotrophs**) are organisms that obtain their energy by eating other organisms, and so all animals are consumers. Consumers that kill and eat other animals are predators, while those that are eaten are called prey.

At each trophic level only about 10% of the energy from the previous level is passed along. The rest of the energy is used by the eaten organism to complete the eight life processes: nutrition, respiration, excretion, response to the surroundings, movement, control of their internal conditions, reproduction, and growth and development.

Apex predator: The final organism in a feeding relationship.

Autotroph: An organism that makes its own food from simple organic compounds in its surroundings, often using energy from light.

Heterotroph: An organism that obtains its food from other organisms; humans are heterotrophs.

Exam tip

You should be able to recall that photosynthetic organisms are the producers of most biomass for life on Earth. It is important you say 'most biomass' and not 'all' because the bacteria at the bottom of the food chains in deep-sea hydrothermal vents are not photosynthetic as it is too dark there.

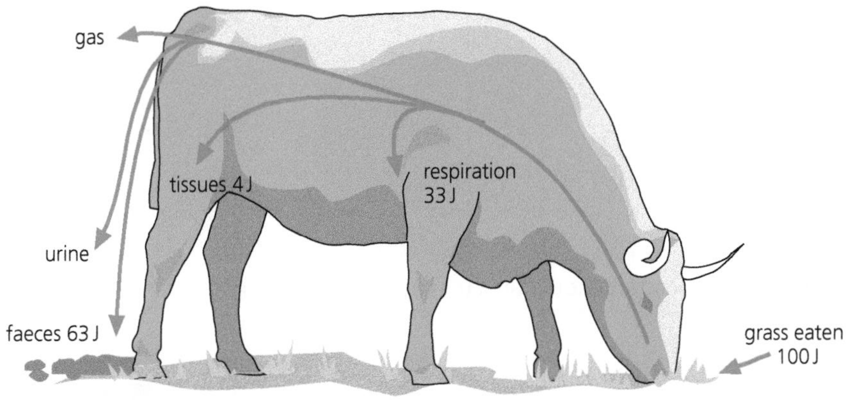

gas

tissues 4 J

respiration 33 J

urine

faeces 63 J

grass eaten 100 J

Figure 6.3 **The energy use by a cow. Compare the food energy eaten with the amount built into body tissue. Look at the large amount of energy lost in faeces. What organisms can use this leftover energy?**

Decomposers

Decomposers break down the remains of dead organisms. This process is called decomposition or rotting. Bacteria and fungi are important decomposers. The speed of decay is slowed by high or low temperatures, as well as by the absence of water and oxygen. This is because decomposing bacteria and fungi find it harder to live in these sorts of conditions.

Gardeners and farmers often try to provide optimum conditions for rapid decomposition of waste material, such as animal dung. These decomposed materials can then be used to produce compost, which is a natural fertiliser for garden plants and crops.

Pyramids of number

REVISED

A **pyramid of number** is a chart that scientists use to show the number of organisms at each trophic level of a food chain. The first trophic level (producer) is lowest in the diagram, with all subsequent trophic levels placed in order immediately above. The width of each bar in the pyramid shows the number of organisms at that trophic level.

> **Pyramid of number:** A graphical way of representing the number of organisms at each level of a feeding relationship.

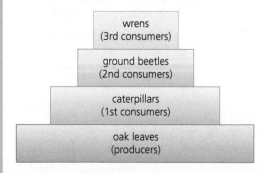

Figure 6.4 A pyramid of number for an oak tree considered in terms of its leaves

Pyramids of number are not always perfect pyramids. Sometimes, if a single large organism like a tree is at the first trophic level, only a small bar might be seen at the bottom. This is shown in Figure 6.5. Equally, if the final trophic level is a parasite like fleas, there will be large numbers of them, and the very top bar could be wider than the one below it.

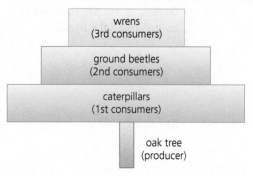

Figure 6.5 A pyramid of number if an oak tree is considered a single organism

Pyramids of biomass

REVISED

Biomass means 'living or recently dead tissue' and it is measured in grams or kilograms. When we measure the biomass of organisms at the different trophic levels of an ecosystem, we see the same sort of pattern as previously.

When represented graphically, this is called a **pyramid of biomass**. In a healthy food chain, the first trophic level will always have the largest mass and it is placed at the bottom of a pyramid of biomass. As you move up each level the mass decreases, which gives the pyramid shape. Unlike pyramids of number, pyramids of biomass in healthy ecosystems are always perfect in that the bars get smaller as you go up the trophic levels. Because the mass of an organism can vary significantly depending upon on the water it has recently consumed, scientists usually measure biomass in dead, dried organisms.

> **Biomass:** Tissue from living or recently dead organisms.
>
> **Pyramid of biomass:** A graphical way of representing the mass of organisms at each level of a feeding relationship.

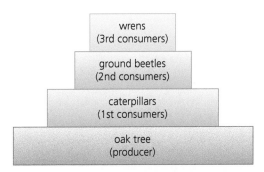

wrens
(3rd consumers)

ground beetles
(2nd consumers)

caterpillars
(1st consumers)

oak tree
(producer)

Figure 6.6 **A pyramid of biomass for an oak tree**

Transfer of energy between trophic levels

Photosynthetic plants and algae are only able to absorb about 1% of the energy transferred by the Sun. However, even this small amount is still enough to power most food chains on our planet. The producers are often eaten by primary consumers, but these consumers are only able to use about 10% of the total energy from the producers. This energy is converted into body tissue, so we can think of energy and biomass as almost interchangeable. Some biomass is lost at each level because:

● Not all of the organisms at each trophic level are eaten.
● Some biomass is lost as faeces.
● Some absorbed material is also lost in other waste (for example, carbon dioxide and water in respiration, or water and urea in urine).

The remaining energy is used by the producers for life processes. Of these processes, the largest amount of glucose is used in respiration.

Pyramids of energy

REVISED

A pyramid of energy shows the total energy present at each trophic level. These pyramids are often very similar in shape to pyramids of biomass. Each higher trophic level should have only around 10% of the energy of the previous level.

Exam tip

You may be asked why the number of organisms is often less at higher trophic levels. You need to remember that this is because only 10% of biomass is transferred between trophic levels, so there are smaller and smaller amounts available the higher you go up the food chain – this is why the levels rarely exceed six in total.

Now test yourself

TESTED

11 What are the stages in a food chain called?
12 What do the arrows in a food chain show?
13 What type of organism is usually found in the first trophic level?
14 At which trophic level can we first see carnivores?
15 What term describes an organism that makes its food from simple organic compounds in its surroundings, often using energy from light?
16 Define the term heterotroph.
17 What do pyramids of number show?
18 What is biomass?
19 What do pyramids of biomass show?
20 Only 10% of energy is transferred between trophic levels. What is the rest of the energy used for?

Answers on p. 128

Exam tip

Exam questions may ask you to draw accurate pyramids of number and biomass from appropriate data. You will be given a section of squared or graph paper to do this. Remember to take your time and draw the bars accurately.

Revision activity

Look up some food chains on the internet and practise drawing both pyramids of number and biomass for them. Check your pyramid of biomass is a 'perfect' pyramid.

Typical mistake

It is important you remember to draw pyramids of number and biomass as rectangles that are touching (as in Figure 6.6). Do not draw them as a pyramid with smooth, diagonal sides.

Exam tip

Exam questions often ask you to describe how biomass is lost between different trophic levels of a food chain. It is important you remember than only approximately 10% is transferred and the rest is used by organisms to complete life processes.

Revision activity

There are a large number of new words in this section. Try making revision cards with the key terms on one side and the definitions on the other. Revisit these cards at a later date to see if you remembered the meanings.

Cycles within ecosystems

All materials in the living world are recycled. This provides the building blocks for future organisms. In other words, some of your atoms will once have been part of previous living organisms. Your atoms will also become part of other organisms in the future.

The carbon cycle

The carbon cycle in Figure 6.7 shows the various compounds that carbon can form and how it is converted between them.

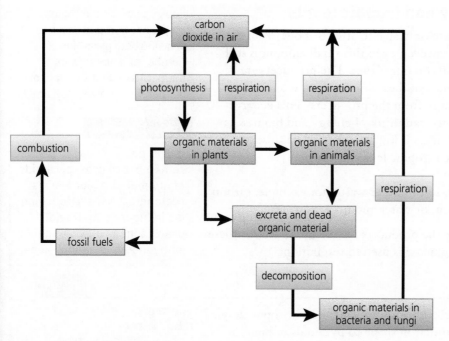

Figure 6.7 The carbon cycle

The four key processes of the carbon cycle are shown in Table 6.2.

Table 6.2 The key processes and conversions of the carbon cycle

Process	Carbon starts as	Carbon ends as
Photosynthesis	Carbon dioxide	Glucose
Respiration	Glucose	Carbon dioxide
Combustion	Fuel	Carbon dioxide
Decomposition	Glucose	Carbon dioxide

> **Revision activity**
>
> To help you revise the carbon cycle, draw the cycle as a series of flow diagrams. Add any extra information alongside the arrows.

> **Combustion:** Burning.
>
> **Decomposition:** Rotting.

The nitrogen cycle

The nitrogen cycle in Figure 6.8 shows the various compounds nitrogen can form and how it is converted between them. Specific bacteria play an important role in this cycle.

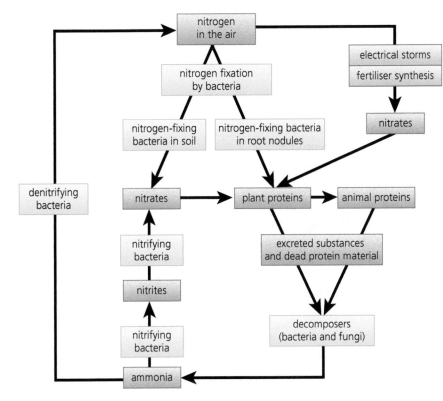

Figure 6.8 **The nitrogen cycle showing the bacteria involved and the changes between nitrogen compounds**

The four key micro-organisms of the nitrogen cycle are shown in Table 6.3.

Table 6.3 **The key bacteria and their conversions in the nitrogen cycle**

Bacterium	Nitrogen conversion
Nitrogen-fixing bacteria	N_2 gas in air to nitrates
Denitrifying bacteria	Ammonia to N_2 gas in air
Nitrifying bacteria	Ammonia to nitrites and then nitrates
Decomposing bacteria	Excreted substances and dead matter to ammonia

Nitrogen-fixing bacteria are often found in root nodules of leguminous plants like peas and beans, as shown in Figure 6.9. Denitrifying bacteria are often found in waterlogged soil.

Now test yourself

TESTED ☐

21 State the four main processes of the carbon cycle.
22 Which process in the carbon cycle converts carbon dioxide to glucose?
23 What compound contains carbon at the start of respiration?
24 Define the term combustion.
25 What are the four main types of bacteria involved in the nitrogen cycle?
26 What conversion do denitrifying bacteria complete in the nitrogen cycle?
27 Where are nitrogen-fixing bacteria often found?
28 Where are denitrifying bacteria often found?
29 What conversion do decomposing bacteria complete in the nitrogen cycle?

Answers on p. 128

Revision activity

To help you revise the nitrogen cycle, draw the cycle as a series of flow diagrams. Add any extra information alongside the arrows, such as the micro-organisms involved.

Exam tip

You need to know the four types of bacteria involved in the nitrogen cycle, but not any of their specific binomial names.

Typical mistake

Exam questions about the carbon cycle usually ask you about its processes, whilst questions about the nitrogen cycle usually ask you about the micro-organisms involved. Many students confuse the two, so make sure you are answering the question being asked.

Figure 6.9 **The roots of a leguminous plant, with root nodules containing nitrogen-fixing bacteria**

Revision activity

Can you explain the carbon cycle and nitrogen cycle to a friend or member of your family? Talking through complex ideas like these will help in your revision.

Human influences on the environment

Air pollution

Air pollution is often caused by waste gases from vehicles or factories. Without suitable regulations, excessive **particulates** also produced from these sources can cause the kind of **smogs** that are sometimes seen covering large cities in some parts of the world.

Waste gases that contribute to air pollution include sulfur dioxide and carbon monoxide. Sulfur dioxide is produced by the burning of fossil fuels and reacts with water vapour to form **acid rain**. This acid rain can destroy entire forests of trees, as well as damage stone buildings and statues. Carbon monoxide is a poisonous gas produced during **incomplete combustion** of fuels. This gas is odourless and can replace oxygen in our red blood cells if we breathe it in. This has the effect of slowly suffocating people.

Lichens are excellent bioindicators of air pollution. These organisms are a combination of a fungus and an alga, which live together in a mutually beneficial relationship.

Particulates: Tiny particles of dust or soot, often produced by burning fossil fuels.

Smog: A thick fog or haze because of smoke or other polluting gases.

Acid rain: Precipitation that is acidic because of a reaction with poisonous gases in the air.

Incomplete combustion: The burning of fuel where there is not sufficient oxygen. This process can produce poisonous carbon monoxide.

The greenhouse effect and global warming

Greenhouse gases include carbon dioxide and methane, as well as water vapour, nitrous oxide and **CFCs**. As humans release larger volumes of greenhouse gases, these gases trap more heat in our atmosphere (hence the term greenhouse). At low levels, the greenhouse effect is needed to keep our planet warm enough to support life. However, analysis of drilled ice cores from the polar regions has shown that human activity in the last few hundred years has doubled the carbon dioxide in our atmosphere.

Human activity that increases the carbon dioxide levels in the atmosphere includes burning fossil fuels in factories, homes and vehicles. It also includes deforestation, which reduces the number of plants absorbing carbon dioxide and so increases the global levels in the air. Deforestation is often accompanied by burning unwanted wood, which of course also increases carbon dioxide levels in the air.

This rapid increase in carbon dioxide is called the enhanced greenhouse effect. It is essential that governments, companies, voluntary organisations and individuals all act to stop this increase because of the damage it is causing.

Global warming is the gradual increase in the Earth's average temperature. The Earth's average temperature has changed naturally over time; for example, we have had ice ages and tropical conditions at various stages in the planet's history. However, almost all scientists agree that the current rate of change is faster than the planet has ever seen before. Scientists also agree that almost all of this increased change is occurring because of the enhanced greenhouse effect.

The consequences of global warming include the melting of glaciers and polar icecaps. This raises ocean levels and threatens low-lying cities like London and New York. Other consequences include freak weather, species migration and threats to **food security**.

Lichen: A symbiotic growth of a fungus and an alga, which are useful bioindicators for clean air.

CFCs (chlorofluorocarbons): CFC molecules used to be widely used in refrigerants and aerosols. Once released, these molecules reacted with the ozone layer in the upper atmosphere and slowly destroyed it.

Food security: How safe the supply of food is.

Exam tip

Exam questions often ask about the cause of global warming (the enhanced greenhouse effect) and what some of the effects are. Make sure you can write a short paragraph explaining these points.

Water pollution

All life needs water. Examples of how humans use water include drinking, growing food, washing and transportation. Pollution of water harms us and the animals and plants that live in, or near, the water that we depend upon. It is estimated that over one billion people do not currently have access to clean water.

Water pollution may come from pathogens like *Salmonella* bacteria, *Norovirus* and parasitic worms. These pathogens are often found in water polluted by sewage. Some parts of the world do not have the same level of toilet facilities that many of us are used to, and the United Nations estimates that the drinking water of 1.8 billion people is contaminated by sewage.

Water can also become polluted by the minerals in fertilisers, which can wash from farmers' fields (known as **leaching**) in heavy rains and enter nearby streams and rivers. This can lead to **eutrophication**. In eutrophication, algae (as the first photosynthetic organisms to benefit from the fertilisers) quickly grow to cover the surface of a pond or lake. This means that no light can reach the aquatic plants, so they soon die. As these aquatic plants rot, the decomposing micro-organisms then use up the remaining available oxygen. This means that other organisms in the water, including fish and aquatic insects, also begin to die.

Some factories still release toxic chemicals into rivers and oceans illegally. Some of these, like the **pesticide** DDT and the metal mercury, cannot be easily excreted as waste. Because these chemicals cannot be excreted, they become locked away in the flesh of organisms. When predators eat this contaminated prey, the chemicals go on to become locked away in them too. This means more and more chemicals concentrate at higher trophic levels. We call this increased concentration of toxins **bioaccumulation**.

Oil spills are another accidental cause of water pollution that can occur. These spills have drastically polluted some coastal areas, killing hundreds of thousands of sea birds, otters and seals.

Typical mistake

The greenhouse effect is an important process that warms the Earth and allows life to survive. The 'enhanced' greenhouse effect is what leads to global warming. Don't get the two confused.

Leaching: The washing of water-soluble plant nutrients from the soil into water, often caused by heavy rain.

Eutrophication: Death of all life in an aquatic ecosystem, often as a result of overuse of fertilisers on nearby farmland.

Pesticides: Chemicals used to kill pests.

Bioaccumulation: The increase in concentration of some toxins at higher trophic levels in a food chain.

Revision activity

Write out the steps in eutrophication as a flow diagram to help you remember it.

Deforestation

Deforestation is the cutting down of trees so that an area can be used for other purposes, often to make farmland for cattle, or rice fields, or to grow biofuels. The first deforestation began around 12 000 years ago as humans began to develop from **hunter-gatherers** to farmers. However, in recent years the rate at which deforestation has occurred has increased massively. Nowadays huge areas of rainforest are often cut down to grow crops like palm oil. Recent figures suggest we have cut down over half of the rainforest that existed 75 years ago.

Deforestation drastically reduces biodiversity because it destroys the habitats of animals and plants. Deforestation also reduces the numbers of photosynthesising plants and so they don't remove carbon dioxide from the atmosphere. This disturbs both the carbon cycle and the balance of atmospheric gases. **Evapotranspiration** is the sum of all water that enters the atmosphere by evaporation and plant transpiration and is a measure of all the water that escapes into our atmosphere. This total is disturbed by deforestation, which means less fresh water enters the atmosphere.

As well as this, the wood from deforestation is often burned, which releases more carbon dioxide into the atmosphere. Deforestation therefore increases the greenhouse effect and global warming.

Hunter-gatherers: Humans who live without farming but by hunting, fishing and collecting wild food.

Evapotranspiration: The sum of all water that enters the atmosphere by evaporation and transpiration.

Revision activity

Draw a mind map of all the content in this chapter. You could use the topic headings: the organism in the environment, feeding relationships, cycles within ecosystems, and human influences on the environment.

Now test yourself

TESTED

30 How is sulfur dioxide released into the atmosphere?
31 What are particulates?
32 Name three greenhouse gases.
33 How do we know there is more carbon dioxide in the atmosphere now than there was thousands of years ago?
34 Define the term greenhouse effect.
35 What are the effects of global warming?
36 What is bioaccumulation?
37 What often causes eutrophication?
38 What is evapotranspiration?
39 How does deforestation affect the enhanced greenhouse effect?

Answers on p. 128

Summary

- A population is the total number of all the organisms of the same species in a geographical area. A community is a group of two or more populations of different species in the same geographical area. An ecosystem is a community of living organisms and the non-living parts of their environment. A habitat is the area or natural environment of a living organism.
- Interdependence refers to how all organisms within a community depend upon each other.
- Sampling is the process of recording a smaller amount of information to make wider conclusions. The numbers and distribution of organisms in an ecosystem can be investigated by using quadrats and transects (imaginary lines along which quadrats are placed). Use of quadrats can be random or systematic.
- Biodiversity is the variety of all the different species of organisms on Earth or within an ecosystem.
- Abiotic factors are non-living (for example light intensity or temperature), whilst biotic factors are living (for example new predators and pathogens). Both abiotic and biotic factors affect the distribution of organisms in an ecosystem.
- Photosynthesising plants and algae are the producers of nearly all biomass (living tissue) on Earth. Food chains and webs show the feeding relationships of organisms within a community.
- Decomposition is the decay of waste material (rotting). The rate of decay is affected by temperature, water and availability of oxygen.
- The number of organisms at each trophic level can be shown in a pyramid of number. The

biomass of organisms at each tropic level can be shown in a pyramid of biomass.
- About 1% of the Sun's light energy is used by photosynthesising organisms. About 10% of this energy is then transferred between each trophic level. The remaining energy is used by the organisms to complete the eight life processes.
- The energy within organisms at each tropic level can be shown in a pyramid of energy.
- Many different materials, including carbon and nitrogen, cycle through the abiotic and biotic parts of an ecosystem. These cycles are important for all living organisms.
- The carbon cycle returns carbon from organisms to the atmosphere as carbon dioxide, which is then used by plants for photosynthesis. Micro-organisms cycle materials through an ecosystem by returning carbon to the atmosphere and mineral ions to the soil during decomposition.
- The nitrogen cycle involves four key types of bacteria: nitrogen-fixing, denitrifying, nitrifying and decomposing.
- Air pollution can be caused by sulfur dioxide and carbon monoxide. Water pollution can be caused by sewage or excess fertilisers. It can lead to eutrophication. Pollution reduces biodiversity.
- Human activities (especially burning fossil fuels) are resulting in an enhanced greenhouse effect, which is causing global warming.
- Deforestation results in disturbance to evapotranspiration and the carbon cycle, and the balance of atmospheric gases.

Exam practice

1 Ecology is the study of living organisms in their environments. Sampling allows us to study a small part of a population and draw conclusions about the whole.
  (a) What are transects? [1]
      A   Square frames of wire used for sampling
      B   Imaginary lines along which sampling occurs
      C   Small containers used to suck up insects
      D   Large nets used to sweep though plants to collect insects
  (b) Which of these is **not** an abiotic factor? [1]
      A   Light intensity
      B   Temperature
      C   Water availability
      D   Disease
  (c) Describe a procedure you could use to investigate whether there are more species of plant on a field where it is cut or left uncut. [6]

2 Feeding relationships of organisms can be represented in diagrams like food chains, pyramids and graphs.
  (a) Define the term pyramid of biomass. [1]

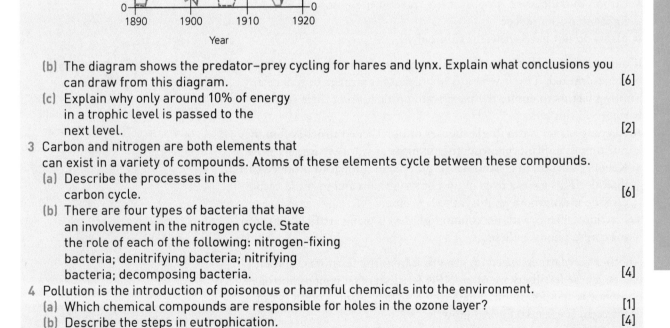

  (b) The diagram shows the predator–prey cycling for hares and lynx. Explain what conclusions you can draw from this diagram. [6]
  (c) Explain why only around 10% of energy in a trophic level is passed to the next level. [2]

3 Carbon and nitrogen are both elements that can exist in a variety of compounds. Atoms of these elements cycle between these compounds.
  (a) Describe the processes in the carbon cycle. [6]
  (b) There are four types of bacteria that have an involvement in the nitrogen cycle. State the role of each of the following: nitrogen-fixing bacteria; denitrifying bacteria; nitrifying bacteria; decomposing bacteria. [4]

4 Pollution is the introduction of poisonous or harmful chemicals into the environment.
  (a) Which chemical compounds are responsible for holes in the ozone layer? [1]
  (b) Describe the steps in eutrophication. [4]

Answers and quick quizzes online

ONLINE

7 Use of biological resources

Using crop plants to produce food

Humans began to evolve from hunter-gatherers to more settled farming communities around 12 000 years ago. Since this time the human population on Earth has increased rapidly to beyond seven billion. To give an idea of how fast the population is increasing, there were fewer than two billion people 100 years ago. This rapid growth in our population and an increase in our standard of living has meant that we are using more and more resources and require ever more food. Now more farms are owned by companies that are under pressure to make money at all costs. This often creates friction between producing ever-increasing yields versus reducing damage to the local environment.

> **Yield**: The amount of an agricultural product produced; for example the yield of an apple farmer is the total mass of apples they produce.

Maximising yield

REVISED

In Chapter 2 we looked at how limiting factors can reduce the rate of photosynthesis in plants and algae (see page 21). When photosynthesis is limited, plants have less available glucose for respiration, growth and repair. Limited photosynthesis therefore means a reduced yield. Limiting factors in photosynthesis occur when:

- temperatures fall (as there is less heat energy present)
- carbon dioxide levels drop, as it is a reactant in photosynthesis
- light intensities reduce
- plants do not have sufficient chlorophyll.

If one or more of these conditions occurs, the rate of photosynthesis becomes limited. This is why it is important for farmers to reduce any limiting factors to ensure the maximum production of their crops. Farmers do this by:

- Keeping plants warm in glasshouses or polythene tunnels, often by burning fuel. Plants photosynthesise more at warmer temperatures.
- Keeping burners in glasshouses or polythene tunnels to produce carbon dioxide. This gas is a reactant in photosynthesis and so more carbon dioxide means more photosynthesis occurs.
- Providing plants with maximum light levels using artificial lights to maximise photosynthesis.

Another way farmers speed up growth is by using fertilisers on their plants. These fertilisers are often NPK fertilisers that contain nitrogen, phosphorus and potassium. These elements help plant growth because:

- nitrogen is essential for leaf growth
- phosphorus is important for development of roots, flowers, seeds and fruit
- potassium is needed for strong stem growth, water movement and the development of flowers and fruit.

Monoculture or crop rotation

Monoculture occurs when farmers choose to grow only one crop. This allows the farm to specialise in one product, meaning the farm can be more effective at planting and harvesting. However, farming only one crop can very quickly lead to nutrient deficiencies in the soil because specific crops often require specific nutrients. This nutrient deficiency means that farmers either need to farm new land or add expensive fertilisers to their soils.

Crop rotation is when farmers grow different crops on rotation in the same fields in different years. This change in crops gives the soil time to recover and means that nutrient deficiencies do not occur. This method of farming is much more traditional, **sustainable** and less damaging than monoculture.

Intensive farming

Intensive farming is a method that maximises farming yields to create the largest financial profit. The following are examples of intensive farming methods:

- Fields are not left **fallow** to recover after harvesting but instead crops are quickly planted again after the previous harvest.
- Larger numbers of animals are raised in confined spaces, which reduces their energy loss to the environment (**factory farming**) and allows them to grow bigger more quickly.
- Fish and other animals are kept in cages to allow them to be easily fed high-protein foods to increase growth.
- Use of chemical fertilisers and pesticides, antibiotics and machines.
- The removal of hedgerows to create larger fields that are easier to manage.

Intensive farming provides much of the food that we eat every day. However, many people now choose to avoid foods produced in this way because they are aware of the damage it can cause to the environment and disapprove of the conditions in which animals can be kept.

Organic farming

In many ways, **organic farming** is the opposite of intensive farming:
- Only natural fertilisers like manure and natural pesticides are used.
- Crops are rotated and fields lie fallow to recover between harvests.
- Animals are kept free range so they can roam more freely.

This method of farming often gives lower yields and requires more work for the farmer. Shops therefore have to charge customers more for organic food than they do for intensively farmed food. Many people are still happy to pay more for organic food because they know the way it has been produced is less damaging for the environment.

Pest control

REVISED

Crop pests reduce yields, meaning less food is produced. Non-organic farmers use pesticides to reduce these pests. These pesticides are often in the form of chemicals that are sprayed onto the crops. Dusting airplanes are sometimes used to do this on a larger scale where there are big fields that need to be treated.

Monoculture: Sustained growth of just one species of crop.

Sustainable: Describes an activity that can continue for long periods without damaging the environment.

Fallow: Leaving an area of farmland empty and without any crops for a period of time to allow nutrients to return to the soil.

Factory farming: Rearing livestock using highly intensive methods.

Organic farming: Non-intensive farming that uses natural fertilisers and pesticides.

Typical mistake

Don't mix up the definitions for intensive farming and factory farming (which is actually a specific example of intensive farming).

Revision activity

Write two lists of the advantages and disadvantages of both intensive and organic farming. Compare and contrast the two lists.

Biological control is another way to kill pests. This method uses a natural enemy or predator of the pest to kill it, as opposed to chemical pesticides. The predator used is a living organism, which is why this method is called biological control. This method is most effective within a glasshouse or polythene tunnel, because movement of the biological control organism is then restricted. Examples of biological control include:

- Ladybird beetles and their larvae are introduced to kill aphids that would otherwise infect roses.
- *Encarsia formosa* is a small wasp that kills whitefly, which would otherwise attack many vegetables.
- Cats are used to kill rodents (perhaps one of the oldest biological control organisms that humans use!).

The advantages and disadvantages of both types of pest control are shown in Table 7.1.

> **Biological control:** The use of a natural enemy or predator of a pest to kill it.

Table 7.1 **The advantages and disadvantages of pest control using pesticides and biological control**

Pest control using pesticides		Pest control using biological control	
Advantages	Disadvantages	Advantages	Disadvantages
Quick and easy to use	Non-specific so can kill other species	Only kills the pest species	Slower as relies upon another species growing
Cheap	Some can bioaccumulate	Cannot bioaccumulate	Can be more expensive
Can be done on a large scale (entire fields)	Pests can become resistant to specific pesticides so that they no longer work	Considered more environmentally friendly	More effective in confined spaces such as a greenhouse (so on a smaller scale)
		Pests cannot become resistant to species used	

Now test yourself

TESTED

1 Define the term yield.
2 State three ways in which farmers maximise photosynthesis in crop plants.
3 What is monoculture and what are its consequences?
4 What is intensive farming?
5 Define the term sustainable.
6 Describe two examples of intensive farming methods.
7 What three elements are often found in fertilisers?
8 Describe two examples of organic farming methods.
9 Define the term biological control.
10 State one example of biological control.

Answers on p. 129

Using micro-organisms to produce food

Yeast

In Chapter 2 we looked at how yeast respires anaerobically in a process called fermentation (see page 31). This process produces ethanol, which we commonly call alcohol. Alcohol is present at about 4% in beer, 12% in wine and 40% in spirits. Yeast is also used to produce bread from dough made of flour and water. As the yeast feeds on the simple sugars in the dough, it produces carbon dioxide, as well as the alcohol, and it is this gas that makes the dough rise. Bread is not alcoholic because the baking process evaporates away any alcohol.

$$\text{glucose} \xrightarrow{\text{energy out}} \text{ethanol} + \text{carbon dioxide}$$

Required practical

Investigate anaerobic respiration by yeast in different conditions

Equipment:
- Flour
- Water
- Sugar
- Yeast
- Plastic cups
- Permanent marker

Method:
- Flour, water, sugar and yeast were added together to make bread dough.
- This dough was divided into equally-sized portions.
- These portions were placed into the bottoms of plastic cups and the height of the dough was marked on the outside of the cups.
- The cups were placed in different temperatures and left for an hour.
- The height on the outside of the cups was recorded and the increase in height was calculated for each temperature.

Results:
- The results showed that the optimum temperature for anaerobic respiration was around 27 °C.
- The results also showed that higher temperatures killed the yeast.
- Lower temperatures slowed the reaction because there was less heat to provide energy.

Yoghurt production

Lactobacillus bacteria are used in yoghurt production. These bacteria break down the sugar (lactose) in milk to form lactic acid. This process lowers the pH of the yoghurt, which helps to preserve it. It also denatures some milk proteins, which gives yoghurt its characteristic sharp taste.

To make yoghurt, milk is first heated to **pasteurise** it. Once cooled, *Lactobacillus* bacteria are added to the milk. The milk is then incubated for four hours at around 40 °C. During this time, the bacteria convert the lactose to lactic acid and so turn the milk to yoghurt. Flavours and colours are then added.

> **Pasteurise:** Heating to partially sterilise milk and make it safe for consumption.

Industrial fermenters

Protein that comes from fungi is called **mycoprotein**. This protein is suitable for vegetarians and has low levels of fat and high levels of fibre. The main brand of mycoprotein in the UK is called Quorn™.

> **Mycoprotein:** Protein that comes from fungi, such as Quorn™.

Mycoprotein is produced from the fungus *Fusarium*. This fungus is grown in large vats called **industrial fermenters**. In this process, fungus is grown on glucose syrup in aerobic conditions inside the fermenter. To maximise the yield of the fungus, no other micro-organisms are allowed to grow inside the fermenter. In fact, contamination by unwanted micro-organisms could even mean the fungus cannot go on to be sold for food. **Aseptic precautions** are therefore used to prevent any infection by undesirable micro-organisms. These precautions include sterilising equipment in alcohol or fire, wearing gloves or other protective equipment, and cleaning up both before and afterwards using antiseptics.

The fermenter is also kept at optimum temperature and pH, with the correct volume of oxygen being added. Finally, paddles or rising bubbles agitate the mixture inside the fermenter to keep it evenly distributed to maximise growth.

> **Industrial fermenters**: Giant containers with regulated conditions that maximise the growth of micro-organisms.
>
> **Aseptic precautions**: A series of techniques used when studying micro-organisms to stop contamination. These techniques include sterilising equipment in alcohol or fire, wearing gloves or other protective equipment, and cleaning before and afterwards using antiseptics.

Now test yourself

TESTED ☐

11 What other name is used for anaerobic respiration in yeast?
12 What are the products of anaerobic respiration in yeast?
13 Why is fermentation important to us?
14 What conversion occurs in yoghurt production?
15 What organism is used to produce yoghurt?
16 Why is milk pasteurised?
17 Which type of organism does mycoprotein come from?
18 What advantages does mycoprotein have over protein from farm animals?
19 What are industrial fermenters?
20 Define the term aseptic precautions.

Answers on p. 129

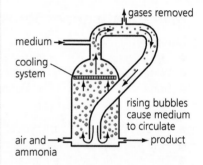

Figure 7.1 **A fermenter for the production of mycoprotein**

Producing food in fish farms

For many years humans thought the stocks of fish in the oceans were limitless. We now think that over 85% of the world's **fisheries** have been overfished. A well-known example of a species that has been overfished is the North Sea cod. Between 1950 and 1980 the mass of cod taken from the North Sea increased by nearly three times. We now understand that we cannot take limitless amounts of fish without consequences. Fish stocks must be maintained at a level where breeding is able to continue or certain species may disappear altogether. Since 1980 humans have taken steps to reduce the number of cod being removed, but numbers have still not returned to 1950 levels.

> **Fisheries**: Area where fish are caught. Fisheries can exist in lakes, rivers, seas and oceans.
>
> **Quota**: A limited or fixed amount that is officially controlled, such as quotas limiting the amount of fish that can be caught.

Some of the steps that have been taken to reduce the number of cod being caught are:
- reducing the lengths of times when boats can fish
- introducing **quotas** that limit the amount of fish that can be caught
- fishing with nets that have larger holes so only older fish that have already reproduced are caught.

Overfishing is a controversial issue because there is conflict between conservation campaigners and local fishing communities who depend upon the fish industry.

Fish farms

Another way of rearing fish without overfishing is using fish farms. Carp, salmon and trout are breeds of fish that are often grown in fish farms. The vast majority of fish grown in farms are for food, because fish are a good source of protein. This farming method often involves breeding fish in tanks, or in enclosures made of net or mesh in existing lakes or seas. If farmers raise fish in small enclosed tanks, they have to aerate (add oxygen to) the tanks, and remove waste by frequently changing the water. If grown in larger net or mesh containers in existing lakes or seas, the water in the farm mixes with that surrounding it. The natural diffusion of oxygen and waste products keeps levels reasonably similar inside and outside the farm.

Even though fish farms are often located in existing bodies of water, farmers still need to check frequently that the water quality is sufficient to ensure the largest possible yield. Farmers also need to ensure that quality food (which does not produce additional harmful waste) is added, either automatically or manually. This again ensures rapid growth and a large yield.

Fish in farms are often kept safe from predation by other species such as birds (otherwise known as **interspecific predation**). However, if fish are carnivorous or the farm is very overcrowded and stressed, the fish may try to eat each other. This is called **intraspecific predation** because it is predation by members of the same species.

Whilst fish farms seem like an excellent way of reducing overfishing, there are potential problems. These include the following:

- Fish are often kept in very close proximity to each other, which can stress them (this is cruel and may also lead to intraspecific predation).
- Disease can spread quickly (so some farmers add antibiotic drugs as a prevention measure).
- Waste produced by the fish can pollute local water courses such as lakes and rivers.
- Habitats can be damaged by the waste produced by the farms.

Selective breeding of fish in farms can help maintain a high quality. You will learn more about this in the next section.

Figure 7.2 **A small fish farm in an enclosure**

Figure 7.3 **A larger fish farm in the sea**

> **Interspecific predation:** The hunting, killing and eating of an organism of one species by an individual belonging to a different species.
>
> **Intraspecific predation:** The hunting, killing and eating of an organism of one species by an individual belonging to the same species.

Now test yourself

21 What steps have been taken to prevent overfishing of cod?
22 What are fish farms?
23 Why are fish kept in farms?
24 What is added to fish farms?
25 If fish are kept in containers, how is their waste removed?
26 What are fish grown in fish farms kept safe from?
27 What type of food group are fish especially good at providing in our diet?
28 What term is given to predation by a different species?
29 What term is given to predation by the same species?
30 How can fish farms in lakes or seas affect the surrounding habitat?

Answers on p. 129

Selective breeding

All dogs are one species, which means that, regardless of their breed, dogs can all interbreed to produce fertile offspring. Humans have used this to their advantage. Genetically, all dogs are descended from wolves, and they came to be domesticated as a result of being selectively bred by humans. Although our ancestors did not understand genetic inheritance or the process of evolution, they did know that if they bred two large dogs, they were likely to get large puppies. We can see that this type of **selective breeding** has occurred repeatedly over thousands of years to result in large breeds like the Great Dane. Similarly, breeding two dogs with protective temperaments over many generations has led to the development of breeds like the German Shepherd. This process of choosing which animals to breed to give desired characteristics is called selective breeding. It is also called **artificial selection**, to remind us that it is a different process from the natural selection that leads to evolution.

Selective breeding has given us Jersey cows, which have been selected to produce creamy milk, as well as Friesian cows, which produce a larger volume of less creamy milk. We have also selectively bred:

- crops that are resistant to disease or give higher yields
- animals that produce more meat
- domestic animals with a gentle nature
- plants with large and unusual flowers.

Almost all crop plants and many of those found in our gardens have been selectively bred. Over many generations the original wild mustard plant has been selectively bred into kale, broccoli, Brussels sprouts, cabbage and cauliflower. These species are different but do have some similarities in their appearance.

Selective breeding can lead to a reduction in the genetic variation of a population. Breeding from closely related individuals is called **inbreeding** and it results in genetic weakening of species due to reduced variation. Occasionally, the selection of a key characteristic (like size) can also accidentally magnify a less desirable one. For example, many pedigree dogs suffer from hip misalignment as a result of selective breeding for another characteristic.

Selective breeding: A process by which humans have chosen organisms to breed together in order to develop desirable characteristics.

Artificial selection: Another term for selective breeding.

Inbreeding: Artificial selection from a small number of parents, which reduces genetic variation.

Typical mistake

Students commonly confuse artificial selection in selective breeding with natural selection in evolution. However, you need to remember that they are not the same thing and what the differences between them are.

Exam tip

Exam questions may ask you to explain the impact of selective breeding in both crops and domesticated animals. Make sure you revise the general process involved, as well as specific examples of each, to help explain your points.

Revision activity

Write a short description of the differences between artificial selection in selective breeding and natural selection in evolution to help you revise.

Now test yourself

TESTED

31 Define the term selective breeding.
32 What is another term you can use instead of selective breeding?
33 How is artificial selection different from natural selection?
34 For what purposes have crops been selectively bred?
35 For what purposes have domestic dogs been selectively bred?
36 What is inbreeding?
37 What are the consequences of inbreeding?
38 Describe a specific example of a difficulty that has arisen as a result of inbreeding.

Answers on p. 129

Genetic modification (genetic engineering)

Genetic modification is a modern technical process by which the genome of an organism is altered by adding a gene from another organism. This allows us to transfer desired characteristics directly into a species. Organisms altered in this way are described as genetically modified (GM) or transgenic. Recombinant DNA is DNA that has been genetically modified, so would not have otherwise been possible in nature.

Genetic modification is much quicker and more precise than selective breeding. Genetic modification is sometimes called genetic engineering. It is an ethical issue and some people disagree with it for religious or moral reasons. Because it is such a contentious subject there are many regulations around genetic modification to tightly control it. For example, it is illegal to genetically modify humans. However, modern medical research is exploring the possibility of using genetic modification to overcome some inherited disorders.

The process of genetic modification

To conduct genetic modification, scientists use restriction enzymes to cut out a specific gene from the genome of one organism. The same restriction enzyme is then used to cut open the DNA of a second organism. The gene is inserted from the first organism into the second organism, and this is then sealed into the DNA with a ligase enzyme.

For example, experiments have been done where the gene that makes jellyfish glow in the dark was inserted into mice. The glow-in-the-dark gene in the jellyfish was cut out with a restriction enzyme, and the same restriction enzyme was used on the mouse embryo. After the glow-in-the-dark gene was sealed into the mouse embryo, this embryo was then implanted back into the uterus of a mouse to grow normally.

It is much harder to insert a gene into every single cell of an adult organism, which is why scientists choose to genetically modify an embryo. This will then divide naturally by mitosis as it grows into an adult organism and so all cells will contain the inserted gene.

Genetically modified crops

Scientists have genetically modified crops to improve food production by making them:
- resistant to disease
- resistant to being eaten by insects or herbivores (examples include cotton, which has been genetically modified to be resistant to an insect pest called the weevil, and soya, which has been modified to be herbicide resistant)
- produce larger yields (for example bigger fruits).

Many people think that genetic modification of crops has the potential to help feed starving people, particularly in countries that experience drought or famine. Other people think that we should not interfere with nature. There are also concerns about the possible spread of genes from genetically modified crops into wild species. It is not yet clear what the effects of GM crops on wild flowers and insects might be. Similarly, it is unclear what the effects of GM crops are on human health, and this is also a concern.

Genetic modification: A scientific technique in which a gene is moved from one species into another.

Transgenic: Describes a genetically modified organism.

Recombinant DNA: Sequences formed by genetic modification that would not have occurred naturally.

Genetic engineering: Another term for genetic modification.

Restriction enzyme: An enzyme used to cut DNA at specific positions. It is used to remove or insert genes in genetic modification.

Ligase enzyme: An enzyme used to seal in cut DNA in genetic modification.

Famine: An extreme shortage of food, often leading to many deaths.

Exam tip

You should be able to describe genetic modification as a process that involves modifying the genome of an organism by introducing a gene from another organism to give a desired characteristic.

Revision activity

Genetic modification is an ethical issue. Research a list of arguments for and against this process in case it comes up in the exam.

Genetically modified micro-organisms

Scientists have also genetically modified bacterial cells, for example to contain the human gene for insulin. These cells are then grown in a large industrial fermenter. As they grow they produce human insulin, which scientists collect to be used to treat diabetes. People with diabetes inject insulin to reduce their blood glucose level because they cannot control it naturally (this is covered in detail in Figure 7.4). Before insulin could be manufactured in this way, people with diabetes used to inject pig insulin. This was not ideal as some people were allergic to pig insulin, whilst others disagreed with it because they were vegetarians or for religious reasons.

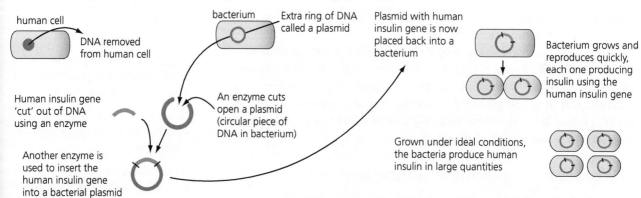

Figure 7.4 Genetic modification of bacteria to produce human insulin

A **plasmid** is often used in the genetic modification of bacterial cells. Plasmids are small, closed, circular pieces of extra DNA found in bacteria. Plasmids can move between bacterial cells and it is often easier to genetically modify a plasmid than the bacterium itself. The gene to be inserted is therefore added to plasmids, which are then taken up by the bacterial cells and carry the new gene into them. The bacteria are now transgenic as the plasmid has acted as a **vector**. Special viruses called bacteriophages are also often used as vectors.

> **Plasmid:** A small, closed circle of DNA found in bacteria, which can move between cells.
>
> **Vector:** A plasmid or virus particle that is genetically modified to carry a desirable gene into another cell.

> **Exam tip**
>
> Exam questions might ask you to explain the potential benefits and risks of genetic modification in agriculture and medicine. To ensure you can answer this question, revise this section carefully and note down an example of each risk and benefit.

> **Revision activity**
>
> Write out the procedure for genetic modification as a flow diagram to help you remember it.

> **Exam tip**
>
> You may be asked to explain the process involved in genetic modification. If you are short of time, you can always write a detailed flow diagram to get the key points down quickly.

Now test yourself

`TESTED`

39 What is genetic modification?
40 What is a transgenic organism?
41 What is recombinant DNA?
42 What do restriction enzymes do?
43 How are ligase enzymes used in genetic modification?
44 Why are fertilised ova or embryos genetically modified instead of entire organisms?
45 Explain why we have genetically modified crops.
46 Why is genetic modification an ethical issue?
47 Which type of organism do plasmids come from?
48 Define the term vector.

Answers on p. 129

Cloning

A **clone** is the genetically identical offspring of a parent that has reproduced asexually. Many plants can reproduce asexually to make clones. Some animals such as water fleas also possess this ability. This is called natural cloning. Clones have no genetic differences from their 'parent' but can show differences that result from **environmental variation**.

Humans can also artificially clone plants and animals. This is used to make genetically identical copies of valuable individuals of a species. For example, Japan is well-known for its cherry trees and it is thought that one especially beautiful tree was cloned thousands of times. Another example of cloning is the banana. Almost every banana that we now eat has been cloned from one first grown in a private estate called Chatsworth House in Derbyshire.

> **Clone:** An organism that is produced asexually and has identical genes to its parent.
>
> **Environmental variation:** Differences in organisms as a result of the environment in which they live.

Cloning plants (micropropagation or tissue culture)

REVISED

We can clone plants by **micropropagation** (also known as tissue culture). During this process, a small number of cells of the parent plant are removed. These cells are placed into a sterile **culture medium** and allowed to grow into a clone of the parent plant. These new clones are called **explants**. We call this growth *in vitro,* which means outside of the body. This process is called **explanting** and it is important in the preservation of rare plants. It also allows plant growers to commercially produce large numbers of genetically identical copies of prize-winning plants.

> **Micropropagation:** Growing a genetically identical clone by removing tissue from a parent plant and placing it into a growth medium. This is also called tissue culture.
>
> **Culture medium:** A liquid or solid substance in which animal or plant samples grow. Also known as growth medium.
>
> **Explant:** A cloned plant that has been grown by micropropagation.
>
> **Explanting:** Transferring living tissue from animals or plants to a culture medium.

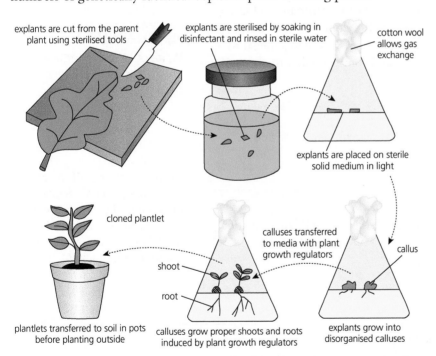

explants are cut from the parent plant using sterilised tools

explants are sterilised by soaking in disinfectant and rinsed in sterile water

cotton wool allows gas exchange

explants are placed on sterile solid medium in light

cloned plantlet

calluses transferred to media with plant growth regulators

callus

shoot

root

plantlets transferred to soil in pots before planting outside

calluses grow proper shoots and roots induced by plant growth regulators

explants grow into disorganised calluses

Figure 7.5 The process of micropropagation

Cloning of mammals

REVISED

In 1996 the first mammal was cloned. This mammal was Dolly the sheep and she was genetically identical to her one parent. This process is shown in Figure 7.6.

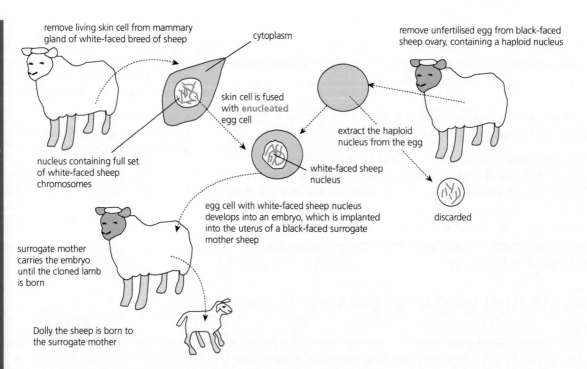

Figure 7.6 **The procedure used to clone a sheep**

It took 434 attempts before Dolly was successfully cloned. She lived until she was six years old, which is less than most other sheep. Some scientists originally thought that this was because she was cloned. However, it has now been proved that her early death was actually because of a lung infection.

Before Dolly, scientists had successfully cloned other animals, including tadpoles and carp fish. Since Dolly we have also cloned various other mammals such as monkeys, pigs, cows and horses. It remains illegal to clone humans. Cloning is an ethical issue and some people disagree with it for religious or moral reasons.

Revision activity

Cloning is an ethical issue. Research a list of arguments for and against this process and make notes.

Typical mistake

Many students confuse the processes involved in genetic modification and cloning. Be careful and make sure you know the difference between them.

Enucleated: Describes a cell that has had its nucleus removed, usually done as part of the process of cloning.

Revision activity

Write out the procedure for cloning as a flow diagram to help you remember it.

Exam tip

Exam questions often ask about the process involved in cloning. If you are short of time, write a detailed flow diagram to get the key points down quickly.

Other forms of cloning

REVISED

Sheep have been genetically modified to produce proteins in their milk, such as blood clotting factors that we use in medicine, and then cloned. This process of genetic modification is shown in Figure 7.7.

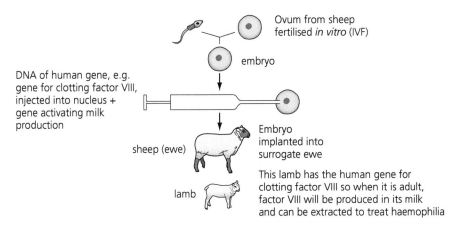

Ovum from sheep
fertilised *in vitro* (IVF)

embryo

DNA of human gene, e.g.
gene for clotting factor VIII,
injected into nucleus +
gene activating milk
production

sheep (ewe)

Embryo
implanted into
surrogate ewe

lamb

This lamb has the human gene for
clotting factor VIII so when it is adult,
factor VIII will be produced in its milk
and can be extracted to treat haemophilia

Figure 7.7 Genetically modified sheep can produce human proteins in their milk.

Another way of cloning is called **embryo splitting**. This process involves splitting a developing embryo, which is then allowed to grow into two identical clones. **Embryo transfer** is the process of putting these embryos back into the mother's uterus to develop normally. Details of these processes are shown in Figure 7.8.

> **Embryo splitting**: The separation of cells of an embryo to increase the number of offspring produced.
>
> **Embryo transfer**: Moving embryos into other animals to increase the number of offspring produced.

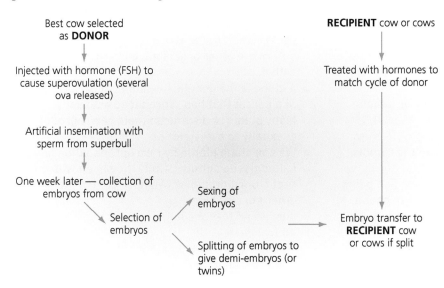

Best cow selected
as **DONOR**

RECIPIENT cow or cows

Injected with hormone (FSH) to
cause superovulation (several
ova released)

Treated with hormones to
match cycle of donor

Artificial insemination with
sperm from superbull

One week later — collection of
embryos from cow

Sexing of
embryos

Selection of
embryos

Splitting of embryos to
give demi-embryos (or
twins)

Embryo transfer to
RECIPIENT cow
or cows if split

Figure 7.8 The processes of embryo transfer and splitting in cows

Revision activity

Draw a mind map of all the content in this chapter. You could use the topic headings: using crop plants to produce food, using micro-organisms to produce food, producing food in fish farms, selective breeding, genetic modification, and cloning.

Now test yourself

TESTED ☐

49 What is a clone?
50 Why can clones sometimes be different?
51 Define the term micropropagation.
52 What other term is used for micropropagation?
53 Why are plants grown by micropropagation?
54 What does *in vitro* mean?
55 How was Dolly the sheep different from every other sheep before her?
56 What have sheep been genetically modified to produce in their milk?
57 Define the term embryo splitting.
58 Why do we undertake embryo splitting and transfer?

Answers on p. 129–30

Summary

- Farmers maximise their yield of crops by keeping plants warm, keeping burners in glasshouses or polythene tunnels to produce carbon dioxide, and providing plants with maximum light levels.
- Intensive farming maximises yields. Examples of intensive farming methods include monoculture, use of fertilisers and pesticides, removal of hedgerows, and keeping animals in smaller spaces. Organic farming is non-intensive and uses natural fertilisers and pesticides.
- Biological control uses another species to control the numbers of a pest species.
- Yeast is a fungus used to make bread and beer. It completes anaerobic respiration or fermentation, which converts sugar to ethanol and carbon dioxide.
- *Lactobacillus* bacteria convert milk into yoghurt by converting lactose to lactic acid.
- Industrial fermenters are large containers with regulated conditions to maximise the growth of micro-organisms.
- Aseptic precautions are a series of techniques used when studying micro-organisms to stop contamination.
- Fish farms raise fish in enclosures or in nets or mesh cages in existing lakes or seas. If small and contained (not in open water), they require the addition of oxygen and the removal of waste.
- Problems with fish farms include the fish being kept in very close proximity, which can cause stress and allow disease to spread quickly. Other difficulties include potential overuse of antibiotic drugs, removal of harmful waste and destruction of local habitats.
- Interspecific predation is the killing and eating of an organism of one species by an individual belonging to a different species. Intraspecific predation is the killing and eating of an organism of one species by an individual belonging to the same species.
- Selective breeding occurs when humans select individual plants and animals to breed in order to give particular characteristics. This can be repeated over many generations.
- Genetic modification is the process of modifying the genome of an organism by the introduction of a gene from another organism to give a desired characteristic.
- It is illegal to genetically modify humans. Genetic modification is an ethical issue.
- In genetic modification, restriction enzymes are used to isolate and cut out the required gene. This is then inserted into the genome of another organism and joined by ligase enzymes. This gene can be directly inserted into the genome of an organism or added to a vector (bacterial plasmid or virus), which then enters the other organism, carrying the gene with it.
- A clone is the genetically identical offspring of a parent that has reproduced asexually. Many plants and some animals can reproduce asexually to make clones.
- We can clone plants by micropropagation (also called tissue culture). Dolly the sheep was the first cloned mammal. We have cloned many other species since.
- It is illegal to clone humans. Cloning is an ethical issue.
- Tissue culture, taking plant cuttings and embryo transplants are all examples of cloning.

Exam practice

1 Industrial fermenters are giant vats in which micro-organisms are grown. Fermenters must be kept sterile using aseptic precautions.

(a) Which of these is not necessarily an example of an aseptic precaution? [1]
- A Flaming equipment
- B Wearing gloves
- C Using antiseptics
- D Growing in agar jelly

(b) Before micro-organisms are cultured in a fermenter, some are genetically modified. Explain why genetic modification is an ethical issue. [1]

(c) Which enzymes are used in genetic modification to join sections of DNA? [1]
- A Carbohydrase
- B Ligase
- C Lipase
- D Restriction

(d) State the word equation for fermentation. [2]

(e) What reaction occurs in order to turn milk into yoghurt? [2]

(f) What conditions are maintained inside a fermenter? [3]

2 Genetic modification, selective breeding and cloning are all techniques used in agriculture to improve livestock.

(a) Use the three terms listed below to copy and complete the table so that they have the correct definitions. [3]

Genetic modification   Selective breeding   Cloning

	Asexual reproduction leading to genetically identical offspring
	Changing the genome of an organism by the introduction of a gene from another organism to give a desired characteristic
	A process by which humans have chosen organisms to breed together to develop desirable characteristics

(b) Describe the steps in selective breeding. [4]

(c) Describe the steps in genetic modification. [6]

(d) Describe the steps that were taken in order to clone Dolly the sheep. [6]

Answers and quick quizzes online

ONLINE

Now test yourself answers

1 Living organisms: variety and common features

1 Energy
2 Respiration
3 To respond to changes in their environment
4 They do not have all eight of the characteristics of living organisms
5 Prokaryotes do not have a nucleus and include bacteria. Eukaryotes do have a nucleus and include animal, plant and fungal cells.
6 Glycogen or fats
7 Fungi or protoctists
8 The way in which some organisms, such as fungi, feed by releasing digestive enzymes outside of their body and absorbing the broken-down food that results from this process.
9 Chitin
10 It contains chloroplasts in which it completes photosynthesis.
11 Micro-organisms (bacteria, fungi or viruses) that cause disease
12 Small, closed circles of DNA found in bacteria
13 A part of a cell with a specific function
14 Organelles, cells, tissues, organs, organ systems, organisms
15 Chloroplasts, cell wall and vacuole
16 Respiration
17 Axon
18 Adult stem cells can only differentiate into one or two types, whereas embryonic stem cells can differentiate into all possible types.
19 Describes a stem cell that can develop into any type of specialised cell
20 Meristem
21 In the tips of roots and shoots
22 Stem cells in plants can usually differentiate throughout the whole of the organism's life. Stem cells usually only have this ability at the start of the organism's life in animals.
23 An issue that some people disagree about for religious or moral reasons
24 A medical procedure in which ova are fertilised outside of a woman, then placed into her uterus to develop into a baby.
25 Fruit and honey
26 Glucose
27 Proteins
28 Amino acids
29 Ribosomes
30 Both are lipids but fats are solid at room temperature and oils are liquid.
31 Iodine solution is added to the food. If it remains brown it contains no starch. If it turns blue or black it contains starch.
32 The enzyme's substrate or substrates
33 The shape of the active site is changed so the substrate no longer fits.
34 The lock and key theory
35 The net movement of particles from an area of higher to lower concentration.
36 In the lungs oxygen diffuses into the blood and then it diffuses from the blood into the body cells.
37 The net movement of water from an area of higher to lower concentration across a partially permeable membrane
38 Hypotonic
39 Osmosis
40 The net movement of particles from an area of lower to higher concentration using energy
41 Energy
42 Diffusion and active transport
43 Minerals are at a low concentration in the soil and a higher one in the plant (so the ions need to move up the concentration gradient).

2 Nutrition and respiration

1 carbon dioxide + water $\xrightarrow{\text{light in}}$ glucose + oxygen

2 $6CO_2 + 6H_2O \xrightarrow{\text{light in}} C_6H_{12}O_6 + 6O_2$

3 Algae
4 It requires energy (provided by light)
5 Palisade mesophyll
6 Temperature, carbon dioxide levels, light intensity and amount of chlorophyll
7 By keeping plants warm in greenhouses or polytunnels, keeping burners in greenhouses or polytunnels to produce carbon dioxide, and providing plants with maximum light levels

8 Anything that reduces the rate of a reaction or stops it

9 Lamp, metre ruler, boiling tube, pondweed and water

10 At the bottom, or the first trophic level

11 All of them (carbohydrates, proteins, lipids, vitamins, minerals, water, fibre), with a correct amount of each food group

12 Fats and oils

13 They require extra food to sustain their baby.

14 Bread, pasta, rice and potatoes

15 It helps with growth of bones and teeth, as well as regulating your heartbeat and helping blood clot.

16 It produces carbohydrase, protease and lipase enzymes, which it releases into the small intestine.

17 The small intestine

18 The rhythmical contraction of the muscles of the digestive system to propel food along it.

19 Carbohydrase, lipase and protease

20 It is produced in the liver and then stored in the gall bladder.

21 glucose + oxygen $\xrightarrow{\text{energy out}}$ carbon dioxide + water

22 $C_6H_{12}O_6 + 6O_2 \xrightarrow{\text{energy out}} 6CO_2 + 6H_2O$

23 It is exothermic because it releases energy.

24 Respiration is always occurring. Photosynthesis is completed only during the day.

25 It is released as heat energy and is used as a chemical store of energy to build larger molecules.

26 In the absence of oxygen

27 glucose $\xrightarrow{\text{energy out (only 5\%)}}$ lactic acid

28 Glucose and oxygen

29 Fermentation

30 glucose $\xrightarrow{\text{energy out}}$ ethanol + carbon dioxide

3 Movement of substances in living organisms

1 Photosynthesis and respiration

2 Respiration

3 carbon dioxide + water $\xrightarrow{\text{light in}}$ glucose + oxygen

4 $C_6H_{12}O_6 + 6O_2 \xrightarrow{\text{energy out}} 6CO_2 + 6H_2O$

5 They absorb water and so the thick inner wall pulls apart.

6 Swollen

7 The point at which the rates of photosynthesis and respiration are equal to one another in a plant

8 Often early to mid-morning and mid- to late afternoon

9 Orange

10 The respiratory system

11 Your neck and abdomen

12 It has rings of cartilage to keep it open at all times.

13 To increase the rate of gas exchange

14 They contract

15 It increases from 0.004% in inhaled air to 4% in exhaled air

16 A medical condition that makes gas exchange difficult as a result of reduced surface area in the lungs

17 Goblet cells and ciliated cells

18 An unhealthy lifestyle, for example a bad diet or smoking

19 Using stents is less invasive than a bypass operation, so recovery is faster.

20 It requires energy to move particles from a lower to higher concentration.

21 The surface area to volume ratio reduces

22 The substances they need can diffuse to all their cells.

23 Xylem cells are dead but phloem cells are alive.

24 Water and mineral ions (to the leaves from the roots)

25 Sugars produced during photosynthesis (from the leaves)

26 Translocation

27 Xylem and phloem

28 To increase the surface area to absorb more water from the soil

29 More wind, less humidity, higher temperatures and light intensities

30 Stomata (also allow guard cells)

31 A potometer

32 Red blood cells, white blood cells, platelets and plasma

33 It gives them a larger surface area to absorb more oxygen.

34 Haemoglobin

35 They do not have a nucleus.

36 They engulf pathogens and destroy them using enzymes.

37 They produce antibodies, which 'clump' pathogens together so phagocytes can destroy them more easily.

38 Antigens

39 They neutralise toxins produced by pathogens.

40 A small quantity of a dead or an inactive form of a pathogen.

41 It means enough organisms in a population are immune that the disease is unlikely to spread, even to those that are not immune.

42 Atria (single: atrium)

43 Valves after the atria and ventricles

44 It has to pump blood further (to the rest of the body).

45 Because blood passes through the heart twice on each circuit

46 An electrical signal made by your natural pacemaker

47 Away from the heart

48 Veins have one-way valves to stop blood under lower pressure from flowing backwards.

49 It increases your heart rate.

50 Tiny blood vessels found between arteries and veins that carry blood into tissues and organs.

51 Pulmonary

4 Coordination and control

1 The removal of harmful waste products from an organism

2 Carbon dioxide and water

3 Oxygen

4 They are all organs of excretion.

5 Renal arteries

6 The nephron

7 Ultrafiltration, selective reabsorption and excretion

8 The collecting ducts

9 Water, glucose, salt ions and urea

10 Urea

11 Anti-diuretic hormone

12 Water loss in urine

13 The maintenance of a constant internal environment

14 To react to changes in your environment and maintain optimum conditions

15 Glucose level in your blood, water level and body temperature

16 The ability of plant stems to grow towards the light

17 The ability of plant roots to grow downwards

18 Two from auxins, gibberellins, ethene

19 Auxins concentrate on the shaded side of the plant. This causes these cells to elongate, which bends the shoot towards the light.

20 The lengthening of specific cells in plants (as a result of hormones)

21 It ripens fruit

22 Gibberellins

23 Any three from: selective weedkillers, rooting powder, tissue culture and fruit ripening

24 When sprayed on weeds in grass, more hormone lands on the larger leaves of the weeds. This causes the plants to grow uncontrollably and die.

25 They encourage cells found in the stems of plants to turn into roots

26 Brain and spinal cord

27 Muscles and glands

28 stimulus $\longrightarrow$ receptor $\longrightarrow$ sensory neurones $\longrightarrow$ motor neurones $\longrightarrow$ effector $\longrightarrow$ response

29 The electrical signal is not initially transferred to the conscious region of the brain.

30 It becomes thinner.

31 When the body becomes too hot, blood vessels near the surface of the skin open to allow more warm blood to flow there to lose excess heat.

32 Hyperthermia

33 Glands

34 The pancreas releases insulin, which travels in the bloodstream to the liver and muscle cells. In these cells the excess glucose is converted into glycogen.

35 Anti-diuretic hormone, which is made in the hypothalamus and stored in the pituitary gland in the brain

5 Reproduction and inheritance

1 Genetically identical organisms (clones)

2 One

3 The asexual reproduction of bacteria

4 Any two of the following:
   • Only one parent is needed.
   • It is a more time- and energy-efficient process as finding a mate is not required.
   • It is faster than sexual reproduction.
   • Many identical offspring can be produced when conditions are favourable.

5 Any two of the following:
   • Production of genetic variation in offspring.
   • If the environment changes, this variation can give a survival advantage by natural selection.
   • Natural selection can be sped up by humans in selective breeding to increase food production.

6 To enclose the flower when it is developing (in bud)

7 The stamen, which is made from the anther and filament

8 The carpel, which is made from the stigma, style and ovary

9 A pollen grain is transferred to the stigma, which starts the growth of a pollen tube down the style until it reaches the ovule inside the ovary. The nucleus of the pollen grain (which

contains the DNA) then moves down the pollen tube and fertilises (joins with) the nucleus of the ovule.

10 By having brightly coloured flowers with a scent and sweet nectar

11 They have smaller, less colourful petals because they do not need to attract insects. Their stamens and stigmas hang outside the flower, which makes it easier for pollen to be blown away from the male anther, and in turn be collected by the female stigma.

12 The spreading of seeds by wind, water, animals or ejection

13 Light, warmth, moisture and the presence of oxygen

14 By forming plantlets on runners

15 Puberty and the menopause

16 Puberty and ovulation

17 14

18 That pregnancy has occurred

19 Oestrogen, progesterone, follicle-stimulating hormone (FSH) and luteinising hormone (LH)

20 Nine months

21 To protect the unborn baby

22 Testosterone

23 Any one of: testes increase in size, voice 'breaks' and becomes deeper, growth of facial hair, or body shape changes (shoulders widen)

24 Any one of: breasts develop, periods begin or body shape changes (hips widen)

25 One copy of all the DNA in your diploid body cells

26 Long, thin structures made from coiled DNA

27 46 (or 23 pairs)

28 A section of a chromosome made from DNA that possesses the code to make a protein

29 Adenine, thymine, cytosine and guanine

30 Weak hydrogen bonds

31 A DNA base linked to a sugar and a phosphate molecule (which make up the backbone of the double helix)

32 The process of making an mRNA copy of a gene sequence of DNA

33 A strand of mRNA attaches to a ribosome and passes through it. For every three mRNA bases, the ribosome attracts a matching transfer RNA (tRNA) molecule with an amino acid attached. A chain of amino acids is produced (a polypeptide), which is then folded into the correct shape to make a protein.

34 A polypeptide is a chain of amino acids that has not yet been folded, whilst a protein is a polypeptide folded into the correct shape.

35 Two copies of the same gene, one from your mother and the other from your father

36 Genotype is the genetic make-up of an organism (represented by letters), and phenotype is the physical characteristics of an organism (as described by words).

37 Heterozygous: A genotype with one dominant and one recessive allele
Homozygous dominant: A genotype with two dominant alleles
Homozygous recessive: A genotype with two recessive alleles

38 25%

39 The interaction of several genes together to control a phenotype

40 XY (or YX)

41 Two genetically identical daughter cells

42 Growth and repair

43 Four non-genetically identical daughter cells (gametes, sex cells, or sperm and ova)

44 To produce gametes (sex cells: sperm, ova and pollen)

45 Environmental factors, genetic factors or environmental and genetic factors together

46 Any suitable examples, such as scars and tattoos

47 Data that come in a range and not in groups

48 Normal distribution

49 A DNA base can be deleted or added. Pairs of bases can be swapped.

50 An organism from which others have evolved.

51 Individual organisms within a particular species show a wide range of variation for a characteristic. Individuals with characteristics most suited to the environment are most likely to survive and breed. These advantageous characteristics are then likely to be passed to the next generation.

52 Fossils are the remains of dead organisms preserved for millions of years in rock. The fossil record shows how organisms have changed over time.

53 MRSA is methicillin-resistant *Staphylococcus aureus*. It is a bacterium that has become resistant to antibiotics in our lifetimes, which is evidence of evolution.

54 The lifecycle of bacteria is much shorter than that of humans so it is easier to see evolutionary changes in them.

6 Ecology and the environment

1 The total number of all the organisms of a species, or the same group of species, that live in a particular geographical area

2 A group of two or more populations of different species that live at the same time in the same geographical area

3 Where all the organisms in a community depend upon each other, and therefore any change to a population of organisms or their environment can affect the whole ecosystem.

4 To save time and money

5 The process of recording a smaller amount of information in order to make wider conclusions

6 In order to count the number of a single species within them.
In order to count the number of different species (a measure of biodiversity).
In order to record the percentage cover of a species, such as grass.

7 If you want to investigate the change across a habitat you would place your quadrats systematically along a transect.

8 A measure of the variety of all the different species of organisms on Earth, or within a particular ecosystem

9 Any two from: light intensity (for photosynthesis), temperature, moisture levels, soil pH, soil mineral content, wind intensity and direction, carbon dioxide levels

10 Any two from: the availability of food, numbers of predators, introduction of a pathogen that causes a communicable disease, competition between species

11 Trophic levels

12 The transfer of energy

13 Photosynthesising plants or algae

14 The third

15 Autotroph

16 An organism that obtains its food from other organisms (for example, humans are heterotrophs)

17 The number of organisms at each trophic level of a food chain

18 Tissue from living or recently dead organisms

19 The amount of biomass at each trophic level of a food chain

20 It is used for the eight life processes.

21 Photosynthesis, respiration, combustion and decomposition

22 Photosynthesis

23 Glucose

24 Burning

25 Nitrogen-fixing, denitrifying, nitrifying and decomposing bacteria

26 Converting ammonia to N_2 gas in the air

27 In the root nodules of leguminous plants like peas and beans

28 In waterlogged soils

29 Converting excreted substances and dead matter into ammonia

30 Through burning fossil fuels

31 Tiny particles of dust or soot, often produced when burning fossil fuels

32 Any three from: carbon dioxide, methane, water vapour, nitrous oxide and CFCs

33 We can see past carbon dioxide levels as recorded in sea ice cores.

34 Where greenhouse gases in the atmosphere trap heat and keep the planet warm

35 The melting of glaciers and polar icecaps, rising ocean levels threatening low-lying cities like London and New York, freak weather patterns, threats to food security and species migration

36 The increase in concentration of toxins at higher trophic levels in a food chain

37 The overuse of fertilisers, often on farmland

38 The sum of all water that enters the atmosphere by evaporation and transpiration

39 It reduces the total photosynthesis that occurs, which increases carbon dioxide levels in the atmosphere. Unwanted wood from deforestation is often burned, which adds more carbon dioxide to the atmosphere. Both of these things act to increase the enhanced greenhouse effect.

7 Use of biological resources

1 The total amount produced of an agricultural product

2 Keeping plants warm in glasshouses or polythene tunnels, keeping burners in glasshouses or polythene tunnels to produce carbon dioxide, and providing plants with maximum light levels

3 Sustained growth of one species of crop, which leads to nutrient deficiencies

4 Industrial agriculture to maximise yield, often involving the use of machines, chemical fertilisers and pesticides

5 An activity that can continue without damaging the environment

6 Any two of: fields are not left fallow to recover but are quickly planted again after the last crop was harvested. Larger numbers of animals are raised in smaller spaces (factory farming). Fish and other animals are kept in cages to allow easy feeding of high protein foods to increase growth. Use of chemical fertilisers and pesticides, antibiotics and machines. Removal of hedgerows to create larger fields that are easier to manage.

7 Nitrogen, phosphorus and potassium (NPK)

8 Any two of: only natural fertilisers and pesticides are used. Crops are rotated and fields lie fallow to recover between crops. Animals are kept free range.

9 The use of a natural enemy or predator of a pest to control its population numbers

10 Any one of: ladybird beetles for aphids, *Encarsia formosa* wasps for whitefly, cats for rodents (or other suitable example)

11 Fermentation

12 Ethanol (alcohol) and carbon dioxide

13 It provides bread and beer.

14 Lactose is converted to lactic acid.

15 *Lactobacillus* bacteria

16 To sterilise it and make it safe for consumption

17 A fungus

18 It is lower in fat and has high levels of fibre.

19 Giant containers with regulated conditions to maximise the growth of micro-organisms.

20 A series of techniques used to stop contamination, which includes sterilising equipment in alcohol or a flame, wearing gloves or other protective equipment, and cleaning before and afterwards using antiseptics

21 Reducing the times that boats can fish. Introducing quotas that limit the amount of fish caught. Fishing with nets with larger holes so as to only catch older fish that have already reproduced.

22 Tanks or enclosures in which fish are grown

23 It increases the yield.

24 Oxygen, if not part of an existing lake or sea, and also food and medicines like antibiotics

25 By changing their water

26 Predation

27 Protein

28 Interspecific

29 Intraspecific

30 Waste from the high density of fish can pollute local water courses.

31 A process by which humans have chosen organisms to breed together to develop desirable characteristics, famously in dogs.

32 Artificial selection

33 Artificial selection is driven by humans and natural selection is driven by nature.

34 To be resistant to disease or have a high yield (or other suitable examples)

35 To have a gentle nature or to be a good guard dog (or any other suitable examples)

36 Artificial selection from a small number of parents, which reduces variation

37 A small gene pool can make inbred organisms susceptible to disease.

38 Many pedigree dogs suffer from hip misalignment (or other suitable example).

39 A process that involves modifying the genome of an organism by introducing a gene from another organism to give a desired characteristic

40 An organism with DNA from another organism (one that has been genetically modified)

41 Sequences formed from genetic modification that would not have occurred naturally.

42 They cut DNA in very specific places.

43 They seal in a section of DNA.

44 It is much easier to insert the DNA into one or a few cells (which then copy the modified DNA as they divide) rather than every cell of an adult organism.

45 To be resistant to disease, to avoid being eaten by insects or herbivores, and to produce larger yields (or other suitable examples)

46 Some people disagree with it for religious or moral reasons.

47 Bacteria

48 A plasmid or virus particle that is genetically modified to carry a desirable gene into another cell.

49 The genetically identical offspring of a parent that has reproduced asexually

50 As a result of environmental variation (not genetic)

51 Removal of tissue from a parent plant and placing it into a growth medium to grow into a genetically identical clone

52 Tissue culture

53 To produce large numbers of genetically identical copies of the best plants

54 Outside of the body

55 She was a clone of her one parent.

56 Proteins such as blood clotting factors

57 The separation of the cells in an embryo to increase the number of offspring produced.

58 Embryo splitting produces clones of desirable organisms and embryo transfer moves these into different animals to develop normally.